37 Fortschritte der chemischen Forschung
Topics in Current Chemistry

Chemical Lasers

Springer-Verlag
Berlin Heidelberg GmbH 1973

ISBN 978-3-540-06099-4   ISBN 978-3-540-38116-7 (eBook)
DOI 10.1007/978-3-540-38116-7

Contents

Chemical Lasers

Dr. Karl L. Kompa

Institut für Anorganische Chemie der Universität München und Max-Planck-Institut
für Plasmaphysik, Garching bei München

Contents

This article deals with a field of research on the borderline between physical chemistry and laser physics. As it is intended to combine aspects of both areas, molecular amplifiers based on partial or total vibrational inversion are first characterized in general, after which the generation, storage, distribution, and transfer of vibrational energy in chemical processes is reviewed. There is a brief discussion of the experimental requirements for laser oscillation and associated hardware problems. Experimental results for specific chemical laser systems are then surveyed and the prospects for high-power chemical laser operation considered. The concluding sections are devoted to the contribution of chemical lasers to reaction kinetics and their other uses in chemistry.

Thus the paper serves three goals: to provide an introduction to chemical lasers; to review the literature up to spring 1972; and to examine some current concepts and perspectives in order to point out possible directions for future developments in this field.

There are two introductory papers on chemical lasers in the literature, both by Russian authors [1]. However, developments in this field are rapid and often divergent so that constant and renewed discussion of this laser concept is called for.

1. Introduction

Chemical Lasers in Physical Chemistry and Technology

Under equilibrium conditions excited molecular states are populated according to the familiar Boltzmann equation, $N_{(excited)} = N'_{(ground)} \exp\text{-}\Delta E/kT$, where ΔE is the excitation energy. For a laser to be possible, the equilibrium has to be disturbed in such a way that a population inversion ΔN arises as defined by Eq. (1) [2].

$$\Delta N = (N - \frac{g}{g'} N') > 0 \tag{1}$$

The prime refers to the lower state, g, g' describe the degeneracy of the states. Under conditions of a population inversion, optical gain is shown on the corresponding molecular transition. Thus, a photon beam passing through the medium at the transition frequency is not absorbed but amplified. The maximum amplification E_l/E_0 for a small input signal E_0 can be written as function of the inversion ΔN, cross-section for stimulated emission σ (see below) and amplifier length l. [2]

$$\frac{E_l}{E_0} = \exp \Delta N \, \sigma \, l \tag{2}$$

There are several ways of creating the required non-equilibrium situation, chemical reactions being one way. Chemical lasers are thus defined as lasers where a population inversion is effected by *selective chanelling of the energy of a chemical reaction* into certain excited product states. We also include in this discussion lasers which are pumped by *energy transfer* from a chemically excited species to an admixture which is then capable of lasing, and lasers

3

which are pumped by *photodissociation*. The following set of reactions (3) illustrates these, three types of pumping schemes:

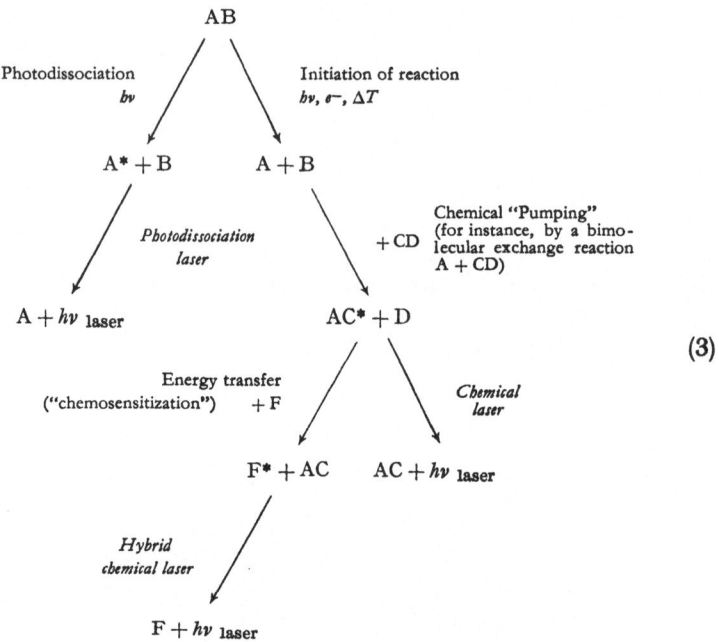

$$(3)$$

A molecule can store energy in the electronic, vibrational, rotational, and translational degrees of freedom. However, the probability that energy can accumulate in these degrees of freedom and can appear in the form of chemical laser emission differs considerably. Fig. 1 shows the usual form of the reaction profile for an exothermic reaction. It is apparent that a product molecule which has just been released from the activated complex is at some distance from its equilibrium state. It contains excess energy which can in principle be given off in two ways, namely by *radiative* or *collisional* processes. There is always competition between these two types of processes. The luminescence quantum yield η_q (4) will be different, depending on the type of excitation.

$$\eta_q = \frac{P_{\text{emission}}}{P_{\text{emission}} + P_{\text{deactivation}}} \tag{4}$$

If the energy $(E_A - \Delta H)$ is sufficient to permit electronic excitation, the probability P of emission is typically $10^6 - 10^9 \ \text{sec}^{-1}$, hence higher than the

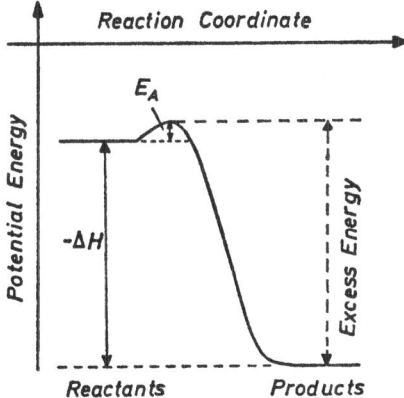

Fig. 1. Typical reaction coordinate for an exothermic reaction

average probability of collisional deactivation. On the other hand, vibrational-rotational transition probabilities are much lower (usually $10-10^3$ sec^{-1}) and the relaxation probability (in relative terms) is correspondingly higher. These figures would favor chemical lasers based on electronic excitation. Unfortunately, electronic excitation is not common in chemical reactions. The energy required for transitions in the optical region $h\nu \leq E_A - \Delta H$ is between 41 kcal/mole (700 nm) and 71 kcal/mole (400 nm). Thus, many simple chemical reactions do not provide enough energy for this type of excitation and those which do may not have products with suitable electronic transitions.

Most chemical reactions proceed via excited vibrational states with energies of $1-10$ kcal/mole associated with one vibrational quantum. This explains why all chemical lasers investigated so far exploit vibrational excitation. In one case purely rotational chemical laser emission has been observed in addition. Only in the related group of photodissociation lasers has emission from electronic states been found. Table 1 shows the various types of pumping processes that are known for chemical lasers today.

As these remarks indicate, chemical lasers employ *infrared chemiluminescence*. As a *method for obtaining kinetic information*, they have to be looked at in relation to other spectroscopic techniques having the same goal. The study of spontaneous vibrational-rotational emission has been most fruitfully applied to fast reactions in the gas phase. This method has experimental limitations due to the relaxation processes competing with spontaneous emission. A very authentic discussion of this method has been given in a recent review by J. C. Polanyi [3]. As opposed to this steady-state technique, chemical lasers permit observations in the pulsed mode.

Since the rate of stimulated emission can be very much faster than the spontaneous emission rate, the time resolution of chemical laser measurements can be quite high. In a very crude way one might say that the rate of emission under lasing conditions can be deliberately increased by a sufficiently intense stimulating field so as to exceed any other collisional rate in

Table 1. Classification of reactions in which chemical laser emission has been found

Type of reaction	Example		
Photodissociation	F_3C-J	$\xrightarrow{h\nu}$	$CF_3 + J$ [a]
Photoelimination	$H_2C=CHF$	$\xrightarrow{h\nu}$	HF [b] $+ HC\equiv CH$
Atom abstraction	$Cl + HJ$	$\longrightarrow$	HCl [b] $+ J$
Chain reaction	$F + H_2$	$\longrightarrow$	$HF^b + H$
	$H + F_2$	$\longrightarrow$	HF [b] $+ F$
β-Elimination following chemical activation	$[F_3C-CH_3]$ [b]	$\longrightarrow$	HF [b] $+ F_2C=CH_2$
Energy transfer ("chemosensitization")	DF [b] $+ CO_2$	$\longrightarrow$	CO_2 [b] $+ DF$

[a] = electronic excitation.
[b] = vibrational excitation.

the system. A more specific discussion of the potential of chemical lasers in reaction kinetics is given in Section 9 of this review.

To assess the role of *chemical lasers in laser physics and engineering*, we have to compare them with other molecular gas lasers. Most high-power gas lasers are pumped either in electric discharges or in supersonic expansions of hot gases (gasdynamic lasers). Both methods offer high powers in the continuous mode of operation and considerable pulse energies in the pulsed mode. It has become clear during the development of chemical lasers that large population inversions can be generated in chemical reactions, too. In addition, there are chemical pumping reactions which are self-sustaining, thus eliminating the need for external energy supplies. The problems, associated with the extraction of power from the reaction system are, however, considerable. A discussion of progress to date will be found in Section 8 of this report.

2. Population Inversion and Molecular Amplification

As stated above, chemical lasers generally originate from non-equilibrium vibrational excitation. This section relates the molecular-state populations with the optical gain to be found on the corresponding transitions. We first envisage a situation of partial equilibrium where different Boltzmann distributions exist in the vibrational and rotational degrees of freedom but there is no communication between the two degrees. This assumption will be justified lateron. The population N_{vJ} of the molecular state with rotational quantum number J and rotational energy $F(J)$ of the v-th vibrational level with density N_v is given for a diatomic molecule as follows: [4]

$$N_{vJ} = \frac{N_v g_J}{Q_J} \exp\left(-\frac{hc}{k}\frac{F(J)}{T_{\text{Rot}}}\right) \qquad (5)$$

(5) assumes a Boltzmann distribution over the rotational states according to the rotational temperature T_{Rot}. If the vibrational states are also defined by a Boltzmann temperature T_{Vib}, Eq. (5) can be rewritten as:

$$N_{vJ} = N\frac{g_J}{Q_v Q_J} \exp\left[-\frac{hc}{k}\left(\frac{G(v)}{T_{\text{Vib}}} + \frac{F(J)}{T_{\text{Rot}}}\right)\right] \qquad (6)$$

Here N is the total number density of molecules, Q_J and Q_v are the rotational and vibrational state sums, and $g_J = 2J+1$ is the statistical weight of the level. Q_J and Q_v are approximated as

$$Q_J = \frac{kT}{h_c B_e}, \quad Q_v = 1 + \sum_{v=1}^{\infty} \exp -\frac{hc}{k} G(v) \qquad (7)$$

The rotational and vibrational energy contributions $F(J)$ and $G(v)$ are given by

$$F(J) = B_e J(J+1), \quad G(v) = (\omega_e v - \omega_e x_e v^2) \approx \omega_e v \qquad (8)$$

with the spectroscopic constants B_e, ω_e, x_e. In Eqs. (5) and (6) T_{Rot} and T_{Vib} are Boltzmann temperatures. In the case considered here, where equilibrium between the degrees of freedom of vibration and rotation is not established, these temperatures are different.

The most important requirement for a laser is that a positive population difference ΔN_{vJ} exists between two vibrational-rotational states.

$$\Delta N_{vJ} = \left(N_{vJ} - \frac{g_J\, N_{v'J'}}{g_{J'}} \right) > 0 \qquad (9)$$

Now we consider a P-branch transition ($\Delta J = +1$ in emission) where $v' = v - 1$, $J' = J + 1$. By substituting Eq. (6) into (9), one obtains the population inversion ΔN_{vJ} for a sample of molecules where both T_{Vib} and T_{Rot} correspond to an equilibrium energy distribution but may still be defined independently:

$$\Delta N_{vJ} = \frac{N\, g_J}{Q_v Q_J}$$
$$\left[\exp - \frac{hc}{k} \left(\frac{G(v)}{T_{Vib}} + \frac{F(J)}{T_{Rot}} \right) - \exp - \frac{hc}{k} \left(\frac{G(v-1)}{T_{Vib}} + \frac{F(J+1)}{T_{Rot}} \right) \right] \qquad (10)$$

Obviously, to make this difference positive the expression in brackets has to be positive. Thus with some rearrangement a limiting condition for the existence of a population inversion is obtained, as first proposed by J. C. Polanyi: [5]

$$\frac{T_{Rot}}{T_{Vib}} < \frac{\Delta F(J)}{\Delta G(v)} \approx 2\,(J+1)\,\frac{B_e}{\omega_e} \qquad (11)$$

Thus, an inversion is always found if the vibrational and rotational temperatures differ such that T_{Vib} exceeds T_{Rot} sufficiently to satisfy this inequality. This is illustrated in Fig. 2 for the case $N_v = N_{v-1}$. The situation here corresponds to an infinite vibrational temperature $T_{Vib} = \infty$. In the extreme case of such an inversion $T_{Rot} \to 0$ and $T_{Vib} > 0$, all the molecules in the various vibrational states are found in the rotational ground state $J = 0$. It can be seen that inversion then exists for the P(1) transitions ($J = 0 \to J = 1$). As this discussion shows, inversion can exist of some J values can only as long as the vibrational temperature is positive ($T_{Vib} < \infty$). This is called a *partial population inversion*. If T_{Vib} attains a negative value, all J transitions may show inversion. This situation, called a *total inversion*, can arise only if $N_v > N_{v-1}$.

T_{Vib} may be only an "effective" temperature, referring solely to the populations of two vibrational levels.

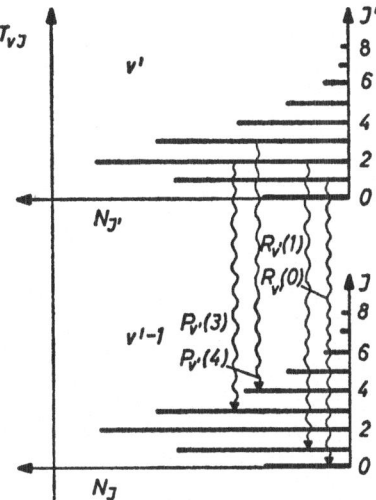

Fig. 2. Partial population inversion between two vibrational levels. The state occupancies N_J and $N_{J'}$ are drawn to scale for HF at $T_{rot} = 300\,°K$ and $T_{vib} = \infty$. Inversion is shown for a pair of high J P-branch and low J R-branch transitions

In (2) the maximal amplification $V = E_l/E_0$ has been introduced as $V = \exp \alpha l$ with the gain coefficient $\alpha = \Delta N_{vJ}\, \sigma$. $\sigma[cm^2] = B_{vJ}\, g(\nu)/\Delta \nu c$ where B_{vJ} is the Einstein coefficient of stimulated emission, $g(\nu)$ is a lineshape factor, $\Delta \nu$ the linewidth and c the speed of light. [2] If these parameters are known, the gain coefficient α can be calculated for any population inversion ΔN. Such model computations are presented in Fig. 3 for a hydrogen-fluoride laser, the most popular chemical laser. [6] It has been assumed for $\Delta \nu$ that the line is Doppler-broadened. The emission probabilities B_{vJ} that have been used have been determined by Talrose and co-workers. [7] One can see from the figure that the gain is always higher for P-branch ($\Delta J = +1$ in emission) than for the corresponding R-branch transitions ($\Delta J = -1$). Partial inversion can lead to gain only in the P branch. As shown in Fig. 4, the line of maximum gain is shifted with increasing rotational temperature to higher-J transitions.

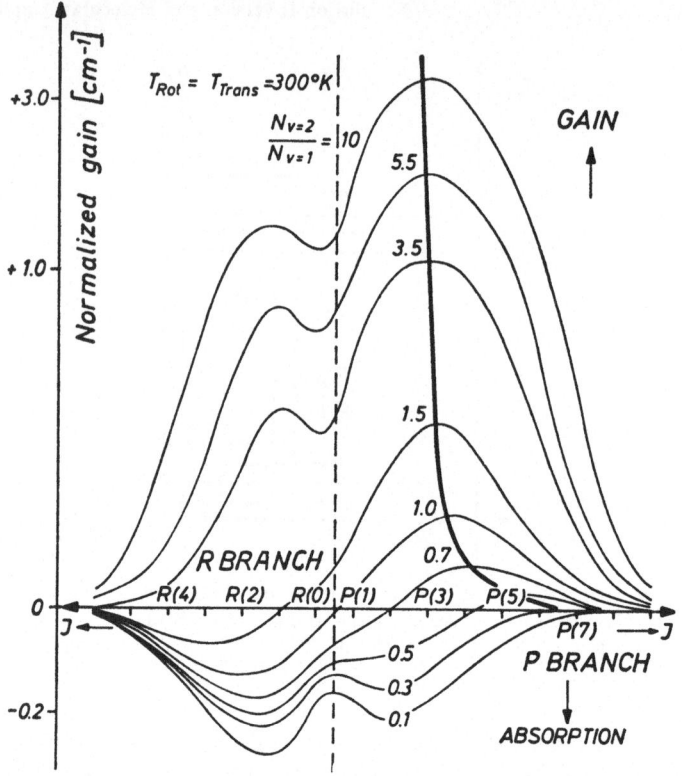

Fig. 3. Relative gain of vibrational-rotational transitions in the hydrogen fluoride molecule for $T_{rot} = 300°$ K and various population ratios $N_{v=2}/N_{v=1}$. It is seen that the gain is always lower in the R-branch than in the corresponding P-branch transitions

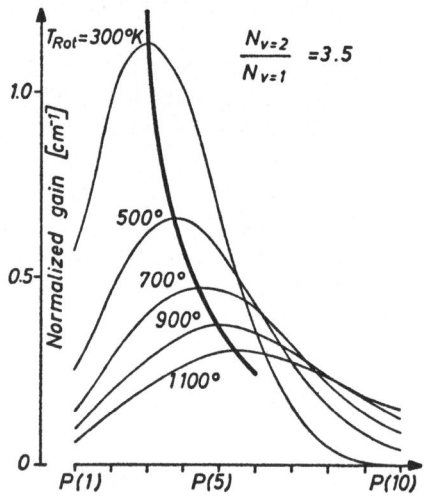

Fig. 4. Shift of the rotational line with maximum gain to higher J by a rise in the rotational temperature. The vibrational inversion is held constant $N_{v=2}/N_{v=1} = 3.5$

10

3. Energy-partitioning in Elementary Chemical Reactions, Vibrational Relaxation

It was shown in the preceding section how optical gain can arise from a partial or total non-equilibrium energy distribution in vibrationally and rotationally excited molecules. This chapter summarizes some experimental facts about the formation and collisional decay of population inversions in simple chemical reactions. The class of reactions most often involved are bimolecular exchange processes, especially the formation of hydrogen halides. Extensive investigations of the energy distribution over vibration, rotation and translation in these reactions have been conducted by J. C. Polanyi and coworkers. Table 2 reproduces some vibrational distribution data obtained from spontaneous infrared luminescence studies. These numbers give some information on the magnitude of the inversion potentially available for use in chemical lasers. Looking, for instance, at the inversion in the reaction $F + H_2 \rightarrow HF + H$ (Table 2), which may serve as a model system in this discussion, one finds the occupancy ratio $N_{v=3} : N_{v=2} : N_{v=1} = 0.48 : 1.0 : 0.31$ with a vibrational energy share E_v of 67% of the total heat of reaction E_{tot}. The reaction preferentially produces hydrogen fluoride HF with $v = 2$. A rough estimate then shows that no more than about 25% of the reaction energy of 34.7 kcal/mole can appear as stimulated radiation (depending on the ground-state population $N_{v=0}$, which cannot be directly measured by this method, and on the rotational temperature T_{Rot}). This figure represents the maximal yield of energy in this process, assuming that F atoms are made available by some means, and ignores the energy necessary for the production of the fluorine atoms.

The gas reactions listed in Table 2 have high rates at room temperature and emission occurs not too far in the infrared. These restrictions are due to limitations of the experimental method which may be overcome in the future. The table could be considerably enlarged by including alkali-metal reactions which have largely been studied by molecular beam methods. [21] Though much discussed, chemical lasers on alkali halides have not yet been realized experimentally. Other results, obtained for instance by flash photolysis/absorption studies, or by the study of combustion, are less detailed and are not included here. But even in this limited form, Table 2 indicates that nonequilibrium distributions which can lead to molecular amplification are often found and are perhaps the rule rather than the exception in simple chemical reactions.

Table 2. Energy distribution and population inversion in reactions of the type A + BC, E_v/E_{tot} = fraction of the total heat of reaction going into product vibration

System studied	Product examined	E_v/E_{tot}	Relative vibrational populations N_v							Refs.
			$v=0$	1	2	3	4	5	6	
H + Cl$_2$	HCl [b]	0.45		0.3	0.6	1.00	0.22	0.03		8,9)
H + Br$_2$	HBr	0.55		0	0.15	1.00	0.98			10)
H + SCl$_2$	HCl	0.43	0.3[a]	0.53	0.72	1.00	0.083	0.25		11)
	HCl	0.35	0[a]	1.0	1.6	1.00	0.4	0.1		12)
Cl + HI	HCl [b]	0.65		0.19	0.37	1.00	0.74			13)
Cl + DI	DCl			0	0.27	1.00	1.9			13)
F + H$_2$	HF [b]	0.67		0.65	2.08	1.00	0	4.55	0.5	14)
	HF [b]		0[a]	0.60	2.00	1.00	0			22,27)
F + CH$_4$	HF	0.67	0[a]	1.45	4.35	1.00	0			22)
	HF [b],c)			1.47	3.8	1.00				17)
F + D$_2$	DF			0.1	0.5	1.00	0.72			15)
F + HCl	HF		0.59[a]	2.95	5.9	1.00	0			16,27)
F + HBr	HF		0.22[a]	0.69	1.11	1.00	0.75			27)
F + HI	HF		0.64[a]	0.79	0.95	1.00	1.21	1.61	1.01	27)
F + H$_2$S	HF [b],c)	0.45		1.0	1.1	1.00	0.75			17)
F + H$_2$O$_2$	HF [b)c)]	0.42		1.09	0.79	1.00	0.47			17)
H + F$_2$	HF			0.32	0.58	1.00	1.18	1.7	1.61	20)
CH$_3$ + CF$_3$	HF			7.7	3.3	1.00	0.23			18)
Hg* + H$_2$CCF$_2$	HF			4.75	1.9	1.00	0.33			18)
H + O$_3$	OH		$v=6$: 0.4; $v=7$: 0.4	7	8; 0.8	9	1.00	14	15	19)
O + CS	CO [d]		≈0.06	0.27	0.66	0.80	0.87	0.64	≈0.2	121)

a) By extrapolation.
b) With other experimental arrangements or in earlier measurements different results were obtained.
c) No emission from SH, OH, O$_2$H, CH$_3$.
d) Infrared emission study, data from other measurements are compared in 9.3.

The relation between the energy distribution in a reaction of the type

$$A + BC \longrightarrow AB + C$$

and the potential energy surface is qualitatively illustrated in Figs. 5 and 6. Two different potential surfaces are shown in Fig. 5. [21]. The approach of atom A to its reaction partner BC may be likened to a point moving along the solid line from the right to the left. The system climbs steadily up the reaction barrier and then descends into the valley which corresponds to the reaction product. For an exothermic reaction, the surface of this valley is lower in energy than the initial system by the amount of the heat of reaction. The difference between Figs. 5a and 5b is that in the first case the new bond A—B is formed slowly and the changes in distance occur smoothly, while in 5b the sudden change in the distance A—B leads to periodic oscillations in the product molecule AB. This is caused by the low activation barrier and the strong attraction of A and B during separation of the products AB and C. A cross-section through the surface, as in Fig. 5, is often pictured as a "reaction coordinate". Fig. 6 shows a diagram based on such coordinates for the process $F + H_2 \rightarrow HF + H$ with HF vibrational excitation according to a reaction surface which might resemble Fig. 5b. The vibrational states which are accessible in this reaction are shown. The k_v's are rate coefficients for the formation of HF at different vibrational levels and determine the vibrational state populations N_v (Table 2) because $k_v \sim N_v$. [22]

Relaxation processes effect decisive changes on the energy distributions as given in Table 2. Such processes usually limit the accumulation and storage of energy in excited states much more than the spontaneous radiation. A rough picture of the relative rates of translational, rotational, and vibrational relaxation is given in Fig. 7. [23] One can see that the redistribution of energy in molecules occurs at varying rates. The energy exchange within one degree of freedom ($V \rightarrow V$, $R \rightarrow R$, $T \rightarrow T$) and between rotation and translation ($R \rightarrow T$) is generally rapid and may be of the same order of magnitude as the reaction rate. The conversion of vibrational into rotational ($V \rightarrow R$) and translational ($V \rightarrow T$) energy is comparatively slower. Thus one can distinguish several stages in the relaxation of a molecular system (12).

The data reproduced in Fig. 7 show that the storage of energy in excited vibrational states is very limited. The total inversion that may be created in a chemical reaction will rapidly decay to a partial inversion with equilibrium within the vibrational degree of freedom but still with a high vibrational temperature $T_{Vib} \gg T_{Rot}$. Laser emission is feasible with both states (I) and (II) of the systems, although with differing degrees of efficiency. In case (I) the stimulated emission rate must be able to compete with the $V \rightarrow V$ transfer rate, while in case (II) it has to compete with the rate of $V \rightarrow T$ transfer. It is apparent from this discussion that a knowledge of

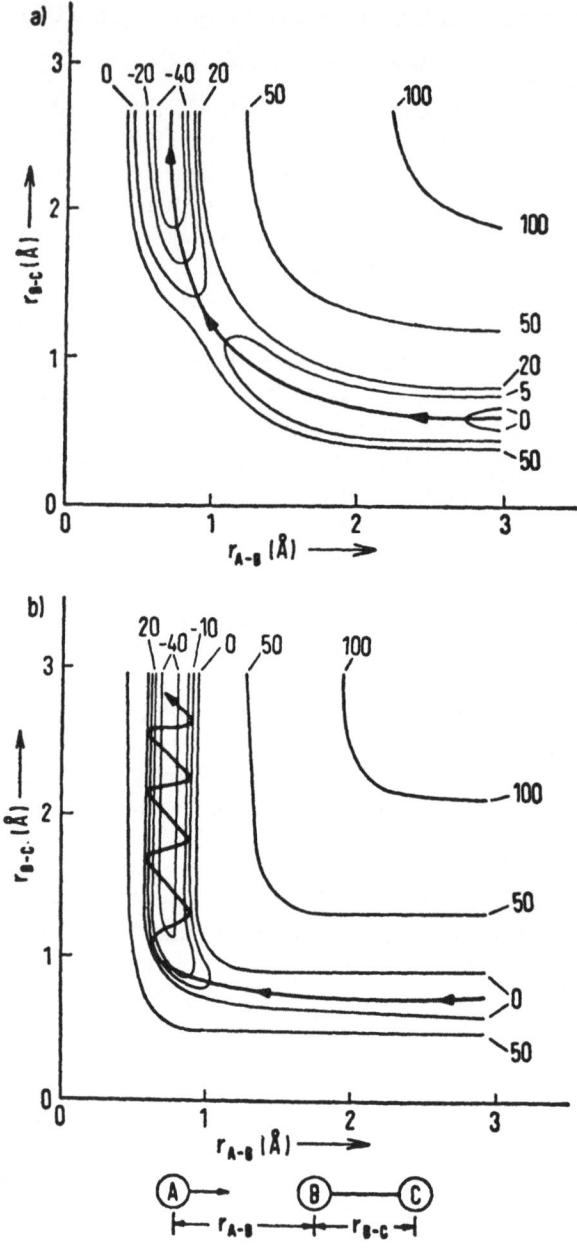

Fig. 5. Potential energy surfaces illustrating qualitatively the formation of vibrationally excited products in reactions of the type $A + BC \rightarrow AB + C$. The influence of the curvature of the reaction path is indicated (after Wagner and Wolfrum[21])

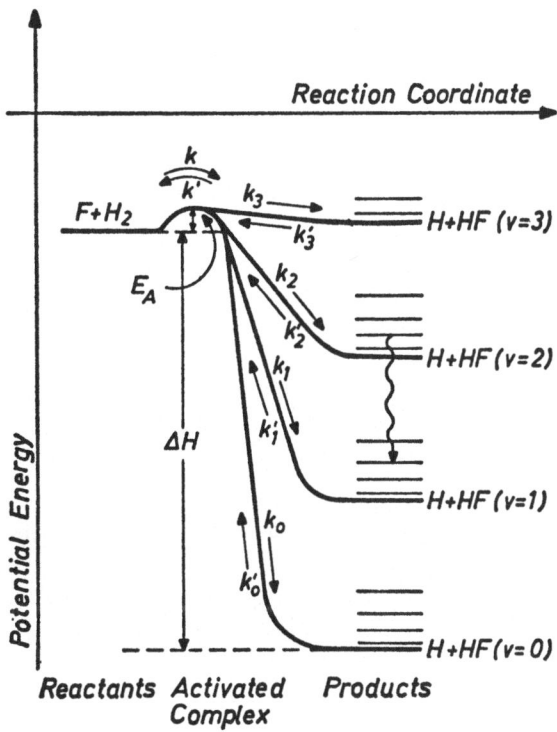

Fig. 6. Reaction coordinates for the formation of hydrogen fluoride in various vibrational states in the reaction $F + H_2 \underset{k'_v}{\overset{k_v}{\rightleftharpoons}} HF(v) + H$

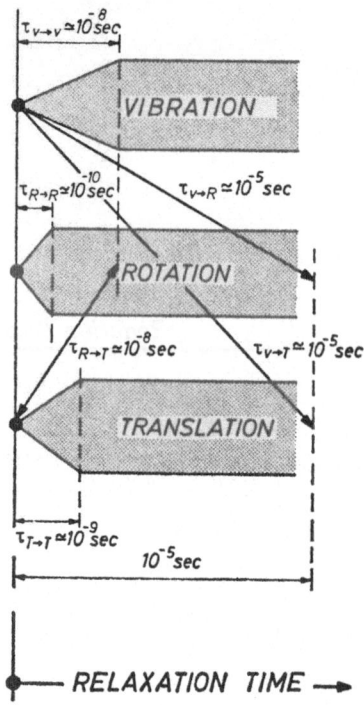

Fig. 7. Scheme of molecular relaxation processes after Flygare[23]. The characteristic times shown are average values for room temperature and atmospheric pressure

$$\downarrow \text{Chemical reaction}$$

(I)   Total inversion

Possible disequilibrium in vibration, rotation, translation

$$\downarrow R \rightarrow R, T \rightarrow T, R \rightarrow T$$

$$\downarrow V \rightarrow V$$

(II)   Partial inversion   (12)

Relaxation within each degree of freedom but still nonequilibrium between vibration and rotation/translation

$$\downarrow V \rightarrow R, T$$

(III)   Total relaxation

16

vibrational relaxation rates is of key importance for the development of chemical lasers. A great deal of relaxation data for small molecules relevant to chemical lasers has been accumulated. Table 3 summarizes some of the results including relaxation by atoms and by molecules. It is seen that the vibrational-translational $(V-T)$ relaxation rate of HF is exceedingly fast. This imposes most severe limitations on the operation of HF lasers. It appears likely that in many of these molecules with small moments of inertia vibrational energy may be transferred into rotational degrees of freedom rather than into translation [61]. The temperature dependence of these rates has been the subject of various investigations.

Table 3. Some HF and HCl deactivation rates obtained by laser-excited infrared fluorescence measurements[a]

Process	Rate constant [sec^{-1} Torr^{-1}]	Ref.
HF $(v = 1)$ + HF $(v = 0)$ ⟶ 2 HF $(v = 0)$	$9.5 \cdot 10^4$	[24]
	$8.7 \cdot 10^4$	[143]
	$6.1 \cdot 10^4$	[147]
	$5.25 \cdot 10^4$	[140]
	$5.2 \cdot 10^4$	[63]
HF $(v = 2)$ + HF $(v = 0)$ ⟶ 2 HF $(v = 1)$	$6.6 \cdot 10^5$	[63]
2 HF $(v = 1)$ ⟶ HF $(v = 2)$ + HF $(v = 0)$	$1.05 \cdot 10^6$	[152]
DF $(v = 1)$ + DF $(v = 0)$ ⟶ 2 DF $(v = 0)$	$2.0 \cdot 10^4$	[140]
	$2.4 \cdot 10^4$	[147]
HCl $(v = 1)$ + HCl $(v = 0)$ ⟶ 2 HCl $(v = 0)$	$8.3 \cdot 10^2$	[148]
DCl $(v = 1)$ + DCl $(v = 0)$ ⟶ 2 DCl $(v = 0)$	$2.5 \cdot 10^2$	[148]
HCl $(v = 3)$ + HCl $(v = 0)$ ⟶ HCl $(v = 2)$ + + HCl $(v = 1)$[b]	$4.25 \cdot 10^3$	[62]
HCl $(v = 1)$ + Cl ⟶ HCl $(v = 0)$ + Cl	$3.5 \cdot 10^5$	[150]

[a] For shock tube studies of these and related processes compare [150], [151].
[b] Nature of process not explicity specified in the reference.

The most popular technique for studying vibrational relaxation phenomena is laser-excited infrared fluorescence. In the commonest case a molecule is resonance-excited to the first vibrational level $(v = 1)$ by a laser source. Thus the relaxation of higher vibrational levels is not easily accessible by this method. Only rarely has excitation of $v > 1$ been achieved, for instance, by direct excitation of overtones [62] or by secondary vibrational exchange processes [63]. Consequently the dependence of the relaxation rates on v is still not clear for most cases.

4. Requirements for Laser Oscillation

If a sample of molecules with inverted populations is placed in an optical cavity which may consist, for instance, of two suitably aligned mirrors (see next section), induced radiation can change the densities N and N' of Eq. (1). The interaction of a system with population inversion with a radiation field of the right frequency ν can be described in the so-called rate equation approximation. The simplified rate equations for the laser are obtained as follows if two states N_1, N_2 are considered with the energy difference $\Delta E = h\nu_{12}$ connected by a radiational transition (13).

Pumping $P_{N_1}(t)$ N_1 (degeneracy g_1)

k_1

$k_1 > k_2$ Spontaneous emission A Absorption B_{21} Stimulated emission B_{12} Deactivation losses $L_{N_1}(t), L_{N_2}(t)$ secondary reactions   (13)

Pumping $P_{N_2}(t)$ N_2 (degeneracy g_2)

k_2

A balance equation for the quantum density q in the laser may be written with regard to the processes shown in (13).

$$\frac{dq}{dt} = A' N_1 + B'_{12} N_1 q - B'_{21} N_2 q - \beta q \tag{14}$$

While $A N_1$ describes the total spontaneous emission rate, $A' N_1$ refers to that part of it that remains in the cavity and contributes to the photon density q. A' is related to the spontaneous emission coefficient A in a more complicated fashion which involves consideration of the resonator modes and the bandwidth of the transition. This will not be discussed in detail here. [2, 25)] The term βq describes the output with the coupling coefficient $\beta = \frac{1}{\tau_c}$ (τ_c = lifetime of the photons in the cavity). $B'_{nm} = B_{nm} \dfrac{g(\nu)}{\Delta \nu}$ where $g(\nu)$ is a line-shape factor and $\Delta \nu$ is the linewidth. The relationship between the Einstein coefficients for spontaneous and stimulated emission

A and B_{12} is given by $\dfrac{A}{B_{12}} = \dfrac{8\,\pi v^2}{c^3}$. Considering now that $g_1\,B_{12} = g_2\,B_{21}$, Eq. (14) may be written as

$$\frac{dq}{dt} = A'\,N_1 + B'_{12}\,q\left(N_1 - \frac{g_1}{g_2}N_2\right) - \frac{q}{\tau_c} \tag{15}$$

Thus the first term in Eq. (15) describes the contribution by spontaneous emission which is important only at the beginning of the oscillation and may be ignored lateron. The second term is the rate of the stimulated processes, while $\dfrac{q}{\tau_c}$ gives the coupling losses. It should be stressed that this type of treatment contains considerable simplifications to allow us to concentrate on the principal features. For a more detailed discussion, reference is made to [2, 25].

The above rate equation for the photons may be supplemented in the same way by equations describing the rate of change of the upper and lower state densities N_1, N_2. The notation here follows from (14) and (15). As before, $\Delta N = \left(N_1 - \dfrac{g_1}{g_2}N_2\right)$ is used and the cross-section for stimulated emission $\sigma\,[\text{cm}^2] = B_{12}\,g(v)\,/\,\Delta vc$ is introduced. The three balance equations for the photons q (compare (15)) and the two state densities N_1, N_2 then read as follows:

$$\frac{dN_1}{dt} = P_{N1}(t) - \sigma c\,\Delta N\,q - L_{N_1}(t)$$

$$\frac{dN_2}{dt} = P_{N2}(t) + \sigma c\,\Delta N\,q - L_{N_2}(t) \tag{16}$$

$$\frac{dq}{dt} = A'\,N_1 + \sigma c\,\Delta N\,q - q/\tau_c$$

It is seen that the excited-state densities N_1, N_2 are determined by a pumping term $P_{N1,\,N2}(t)$ and a loss therm $L_{N1,\,N2}(t)$ referring to the collisional processes. The loss rate $L_{N1}(t)$ might, for instance, be given by $\dfrac{dN_1}{dt} = k_q\,N_1\,M$ where k_q is a quenching-rate coefficient and M is the concentration of particles effective in deactivation. In addition, N_1 is decreased by the stimulated processes which in the same way increase the lower-state density N_2. The competition between these two types of losses for N_1 is evident. With values for the cross-section σ of $\sim 10^{-16} - 10^{-18}$ [cm^2] the rate constant $B'_{21} = \sigma c$ becomes $\sim 10^{-6} - 10^{-8}$ [cm^3 sec^{-1}] hence far larger than any rate constants for bimolecular collisional processes. For purposes

19

of comparison, it may be mentioned that the very fast $V \to T$ transfer rate of hydrogen fluoride, discussed in Section 3, has a rate constant of the order of $10^{-11} - 10^{-12}$ [cm^3 sec^{-1}]. Thus it should be stressed that with a suitable quantum density q and inversion ΔN the stimulated emission rate can easily exceed the rate of any collisional deactivation.

Let us now ask what is the value of the inversion ΔN which effects the stimulated emission rate. The inversion in a laser cavity cannot build up to more than a threshold inversion ΔN_0 which can be obtained from the threshold condition of Schawlow and Townes.

$$R_1 R_2 T^2 V^2 = 1 \tag{17}$$

Here V is the amplification for one traversal and V^2 for a round trip of the photons in the cavity with mirror-loss coefficients R_1, R_2 and transmission loss T. The optical gain V on the other hand is described by Eq. (2) as $V = \exp \sigma \Delta N l$. With Eqs. (2) and (17) the threshold inversion density ΔN_0 is obtained.

$$\Delta N_0 l = \frac{1}{2\sigma} ln (R_1 R_2 T^2)^{-1} \tag{18}$$

So with the values of $\sigma \sim 10^{-16} - 10^{-18}$ quoted above and a total loss coefficient $R_1 R_2 T^2 = 0.3$, the total threshold inversion is $\Delta N_0 l \sim 10^{16} - 10^{18}$ [cm^{-2}]. For the somewhat idealized conditions of a photochemical iodine laser (Chapter 6.1) the balance equations (16) have been solved numerically in [25]. If a fast pumping rate $P_1(t)$ is assumed, the change of the inversion and quantum density might appear as shown in Fig. 8. After threshold is reached, the inversion ΔN starts dumped oscillations around the threshold inversion ΔN_0 and later reaches stationary conditions. Thus, the part of the inversion ΔN which is available for laser output is that in excess of ΔN_0.

$$\Delta N' = \Delta N - \Delta N_0 \tag{19}$$

The maximum energy that can be expected in the laser pulse is

$$E_{\max} = h \nu (1 + g_1/g_2)^{-1} \Delta N' \tag{20}$$

It is clear from Fig. 8 how important it is that the threshold of oscillation be reached very fast to enable the stimulated processes to compete with the collisional deactivation processes. It has been assumed so far that the full inversion under the entire spontaneous-emission line profile contributes to

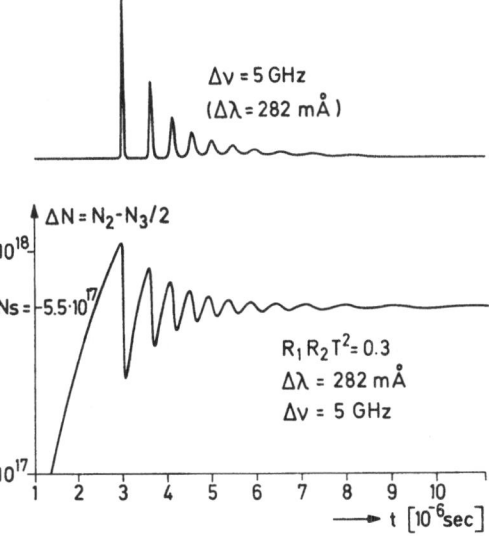

Fig. 8. Calculated time-dependent quantum density (upper plot) and corresponding inversion (lower plot) of a photochemical iodine laser with given linewidth and resonator parameters. After threshold is reached the inversion exhibits oscillations and shows a steady-state behavior later on

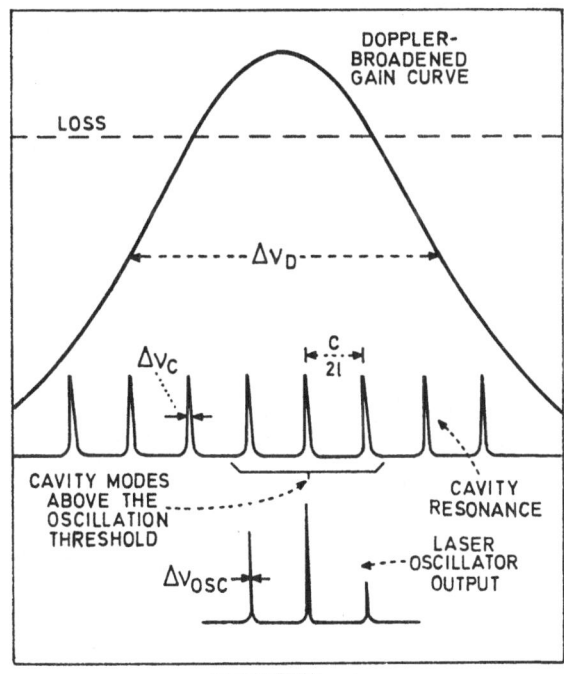

Fig. 9. Operation of a gas laser in several axial cavity modes spaced by $(c/2l)$ Hz (off-axis modes are not shown in the figure). The three cavity modes near the peak of the gain profile are above threshold for oscillation[65]

laser oscillation. However, oscillation is possible only at frequencies corresponding to cavity resonances. These cavity modes spaced by $c/2\,l$ are usually much narrower than the emission line, as is apparent from Fig. 9. The requirement then is that during the period of laser oscillation some mechanism must operate to change the energy of the molecules so as to bring them into the cavity modes. This requirement is normally met by collisions of the excited molecules and is called homogeneous broadening of the laser transition. It is easily achieved in gas lasers at moderate pressures and for not too short a pulse duration. Only then, however, can quantitative information on the inversion number densities be obtained from chemical laser experiments (see Section 9).

5. Design Parameters of Pulsed Chemical Lasers

Three steps may be distinguished in the operation of a pulsed chemical laser:

1) the initiation of the chemical pumping process

2) the formation (and subsequent partial decay) of the inversion in the reaction

3) the extraction of power from the reaction system.

The first step involves the preparation of the reactants by some energy input from an external source. There are several ways of doing this. The earliest chemical lasers were pumped in flash-photolysis experiments and this is still a very popular method. Now, however, the initiation of the pumping reaction in an electric discharge is gaining in importance since transverse discharge geometries have proved to be very convenient for the excitation of other molecular lasers, for instance, CO_2. Comparing the two methods, it may be said that flash photolysis provides somewhat, "cleaner" conditions since the absorption of molecules, photon absorption cross-sections, and photochemical effects are in most cases well known. The laser reaction can be started in a controlled fashion by selectively dissociating one reaction partner. On the other hand, a discharge can provide a more efficient energy input and usually allows a higher pulse-repetition rate, if desired.

Other sources of energy input that have been used are dissociation by shock waves and by high-energy electron beams. A further possibility which has been discussed but not realized so far is to supply the energy by means of another laser, preferably in the infrared [64].

Two experimental set-ups are described here. Both are used in the author's laboratory but represent general design features. An important requirement, especially for hydrogen-halide lasers, is a fast excitation pulse and a high pumping rate to reduce the effect of vibrational deactivation. Here fast coaxial flashlamp-laser arrangements of the type developed originally for the excitation of dye lasers have proved most useful [26]. A small chemical laser of this kind is illustrated in Fig. 10a, the design of the lamp being shown schematically in Fig. 10b. The light pulse from a xenon

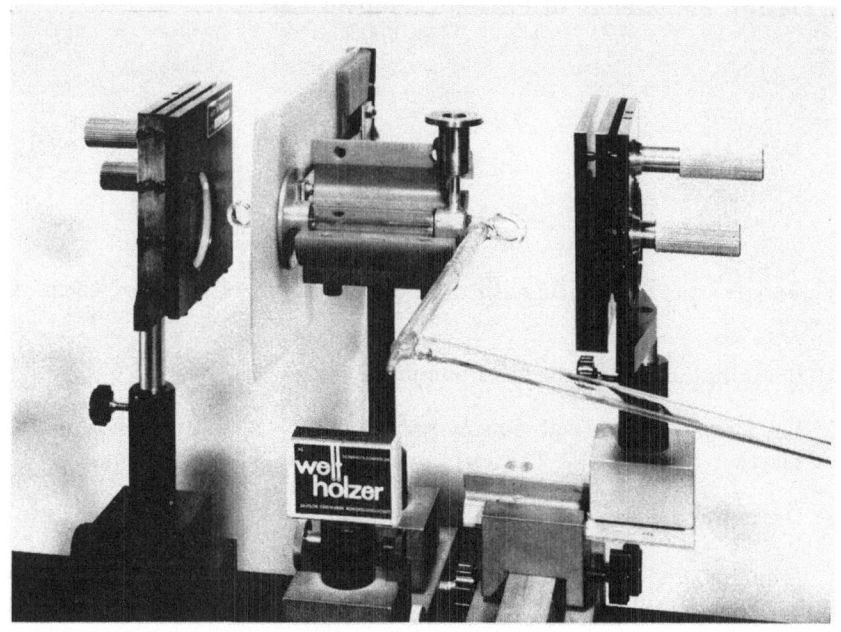

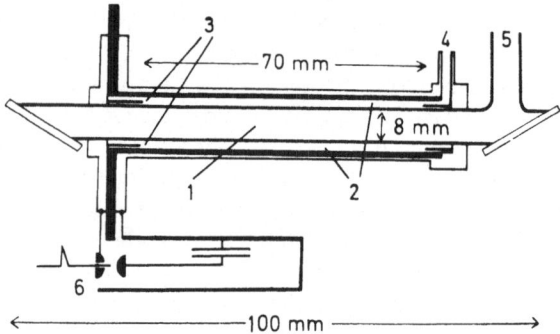

Fig. 10. Laser oscillator with coaxial flashlamps. *1* laser tube suprasil, *2* flashlamp, *3* tungsten electrodes, *4* filling port for flashlamp, *5* to vacuum line, *6* high voltage pulse capacitor with switching gap[25]

lamp of this design but with a larger illuminated length had a risetime and halfpeak duration of .5 μsec with a discharge energy of 1.2 kJ. Actinometric data are available for some flash-lamps [28]. A diagram of a typical flash-photolysis laser experiment is given in Fig. 11.[27]

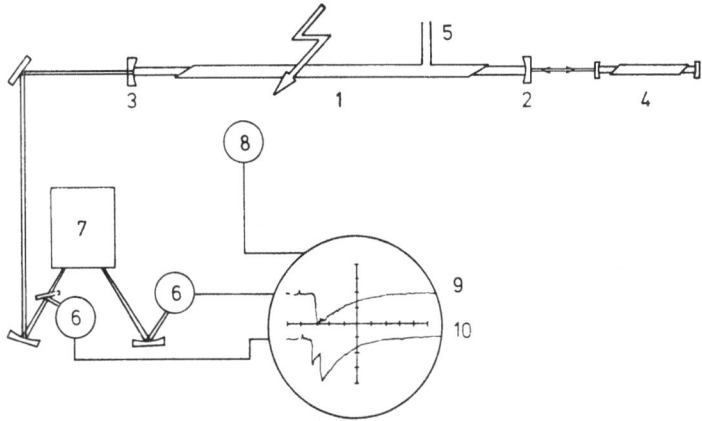

Fig. 11. Chemical laser arrangement with flash initiation. *1* laser tube, *2* 100% reflecting cavity mirror, *3* coupling mirror, *4* He-Ne laser for resonator alignment, *5* to gas manifold, *6* infrared detectors, *7* monochromator, *8* diode providing a triggering signal for the oscilloscope, *9* single spectral line, *10* total emission signal [6]

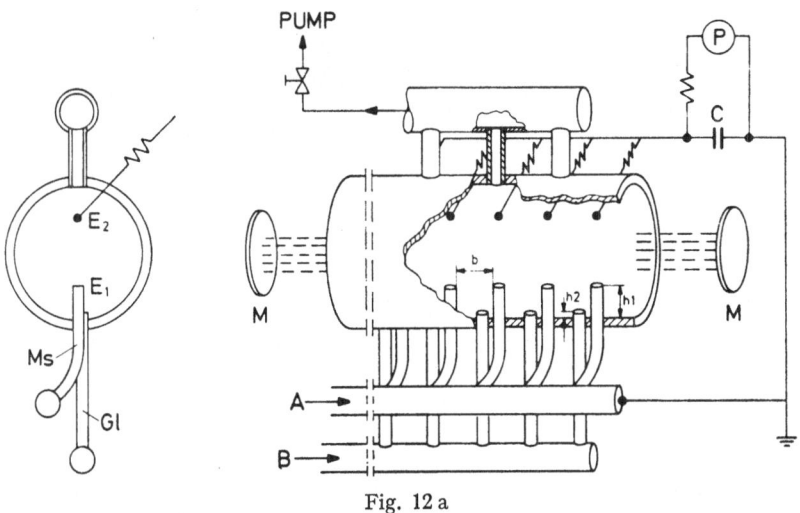

Fig. 12 a

Fig. 12 a and b. Chemical laser with flow of chemicals transverse to the resonator axis and transverse discharge geometry. $E_{1,2}$ electrodes, E_1 (brass) serving as the inlet tube for one of the reagents fed by A. B feeding line (glass) for the other reagent. The relative heights $h_{1,2}$ of the inlet pipes can be varied. M cavity mirrors. P,C pulse discharge unit [39]

25

Fig. 12 b

In the other experimental apparatus the flow of chemicals and the discharge are both transverse to the cavity axis. Details are shown in Figs. 12a and 12b.[29] Improvements are possible if the single row of pin electrodes is replaced by a brush or shower discharge with many pins in a two-dimensional arrangement. The discharge then fills a larger volume, as can be seen in Fig. 13.[30] The prime experimental problem is to achieve a homogeneous electron distribution over a large laser-tube cross-section. Discharge techniques developed recently for CO_2 lasers [31], for instance, preionization or electron-beam initiation, might now be considered for application in chemical lasers. However, we need to know more about the dissociation and excitation by electron impact of molecules used in chemical lasers in order to design experiments so as to ensure optimal distribution of the energy of the electrons.

Fig. 13. Pulse discharge-initiated laser with transverse-discharge geometry. Top and bottom lifted to show the multiple-pin electrodes. The trough above the pins is filled with an electrolyte solution serving as a resistor for electrically decoupling the electrodes[30]

6. Specific Chemical Laser Systems

The idea of the chemical laser is nearly as old as the whole area of experimental laser physics. The first meeting on the subject, in 1963 [32], was organized by the American Optical Society, and chemical laser emission was reported for the first time in 1965 by Kasper and Pimentel [33]. The emission occurred in the photolytically initiated hydrogen-chlorine explosion ($H_2/Cl_2 \rightarrow 2\ HCl$). Predating this discovery, the first photodissociation laser was described by the same authors in 1964 [34]. This laser was based on the formation of excited iodine in the photochemical dissociation of alkyl iodides, preferentially trifluoromethyl iodide.

Interest in the physics and technical applications of chemical lasers appeared somewhat later. Unlike other types of laser — gas lasers and solid-state lasers — which soon found various useful applications, chemical lasers remained for many years within the field of physical-chemical research. The reason for this may well be that, before progress could be made in either the theory or the applications of chemical lasers, certain concepts from various disciplines had to be combined. However, from about 1969 an increasing number of technically oriented papers on chemical lasers have appeared in the literature. Continuous lasers were of course, developed first (Section 6.7), but the interest in pulsed chemical lasers is now growing.

6.1. Photodissociation Lasers

Chemical lasers are pumped by reactive processes, whereas in photodissociation lasers the selective excitation of certain states and the population inversion are directly related to the decomposition of an electronically excited molecule. Photolysis has been the only source of energy input employed in dissociation lasers, although it appears quite feasible to use other energy sources, *e.g.* electrons, to generate excited states. Table 4 lists the chemical systems where photolysis produces laser action. It is appropriate to begin the discussion of Table 4 with the alkali-metal lasers since Schawlow and Townes in 1958 [35] chose the $5\ p \rightarrow 3\ d$ transitions of potassium for a first numerical illustration of the feasibility of optical amplification. These historical predictions were confirmed in 1971 by the experimental demonstration of laser action in atomic potassium, rubidium and cesium (Fig. 14).

Table 4. Photodissociation lasers

Photolysis	Active species	Refs.
$CF_3I \xrightarrow{h\nu}$	$CF_3 + I(5\,{}^2P_{1/2})$	25,34,36–54)
$NOCl \xrightarrow{h\nu}$	$Cl + NO(X^2\,\Pi_{1/2,\,3/2}\,v > 0)$	55,56)
$C_2N_2 \xrightarrow{h\nu}$	$2\,CN(X^2\,\Sigma,\,v.> 0)$	57)
$IBr \xrightarrow{h\nu}$	$I + Br(4\,{}^2P_{1/2})$	58)
$K_2 \xrightarrow{h\nu}$	$2\,K(5_p\,{}^2P_{1/2,\,3/2})$	59)
$Rb_2 \xrightarrow{h\nu}$	$2\,Rb(6_p\,{}^2P_{1/2,\,3/2})$	59)
$Cs_2 \xrightarrow{h\nu}$	$2\,Cs(7_p\,{}^2P_{1/2,\,2/3})$	59)
$CF_3Br \xrightarrow{h\nu}$	$CF_3 + Br(4\,{}^2P_{1/2})$	60)

The pump sources here were various giant-pulse lasers. In the case of cesium, emission of a variety of transitions is observed by cascading from the $7p$ via the $5d$ states down to the $6p$ states and from the $7p \to 7s$ transition. Fig. 14 shows the possible cesium laser lines.

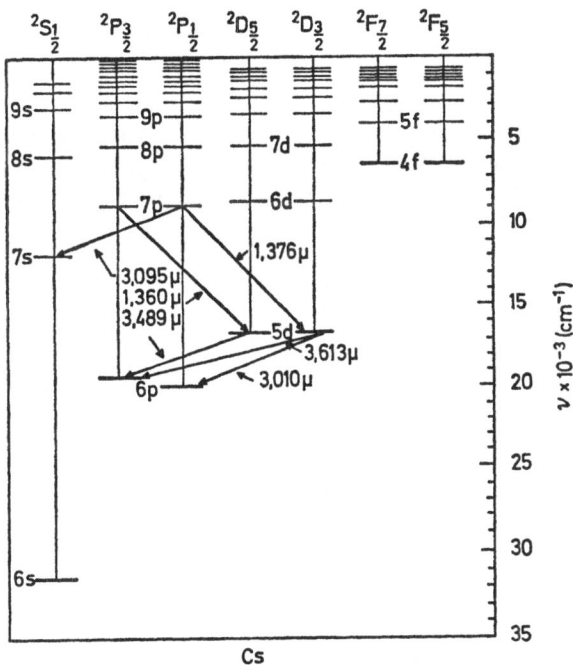

Fig. 14. Spectroscopy of the cesium photodissociation laser[59]

Among the other photodissociative laser systems listed in Table 4, the NOCl and IBr photolyses have received attention because of their chemical reversibility. With IBr in particular, the starting compound is recovered within a few milliseconds, thus allowing a repetition rate of some 10^2 Hz without the need for a vacuum system for refilling the laser tube. Although the C_2N_2 photodissociation is reversible, too, photolysis is required in the vacuum UV and only very weak emission signals have been reported. Photolysis in the vacuum UV with $\lambda < 2000$ Å is also necessary for the CF_3Br dissociation.

Much the most carefully investigated photodissociation laser is the iodine laser from the photodissociation of alkyl iodides, preferentially CF_3I or $i-C_3F_7I$. This laser, reported first by Kasper and Pimentel in 1964, uses the excitation of iodine to the $5^2P_{1/2}$ state in the photolysis of the iodides according to the following much simplified scheme.

$$\text{Photolysis} \quad CF_3I + h\nu_{2700\,\text{Å}} \longrightarrow CF_3 + I(^2P_{1/2})$$

$$(21)$$

$$\text{Laser} \quad I(^2P_{1/2}) \longrightarrow I(^2P_{3/2}) + h\nu_{13150\,\text{Å}}$$

Breaking the CF_3I bond requires ~ 2.5 eV; photolysis is accomplished via an absorption band of the CF_3I molecule at around 2700 Å (~ 4.75 eV) [25] which gives ample excess energy for exciting the products. A variety of secondary chemical reactions and collisional deactivation processes changes the concentrations of I* and I during and after the photolysis flash. A rough picture of the relevant processes is given in the level scheme of Fig. 15.

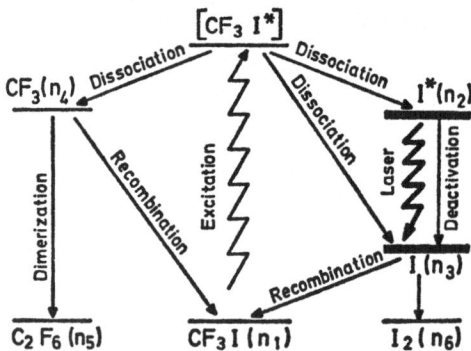

Fig. 15. Level scheme of the photochemical iodine laser. The primary process of photo-dissociation as a result of molecular excitation and some of the secondary processes are indicated

One can see that the final photolysis products are hexafluoro ethane and molecular iodine. The rate constants for these processes have either been measured or can at least be estimated with sufficient accuracy, so that we can make the following classification with regard to their influence on the I* concentration. Detailed numerical calculations have been carried out for the set of rate equations shown above. The experimental conditions were specified so that pressures up to 100 Torr of CF_3I were used in a laser tube of $\sim$1 cm i.d. The light input by the flash-lamp was measured as $\sim 10^{17}$ quanta/cm^3. [25)] This corresponds to 9 J of useful photolysis light in the range of the absorption band of CF_3I centered around 2700 Å. Flash duration (half-peak) was 4 μsec. The computations then confirm that two time domains can be distinguished. Shortly after beginning of the photolysis the two most important chemical reactions are the dimerization of CF_3 radicals (Reaction 9 in Eq. (22)) and the recombination of ground-state I and CF_3 to regenerate CF_3I (Reaction 8). Thus two pumping processes have to be considered: the primary photolysis to give I* and the removal of I by

1.	$I^* + CF_3$	$\longrightarrow$	$CF_3 + I$	O	
2.	$I^* + C_2F_6$	$\longrightarrow$	$C_2F_6 + I$	+	
3.	$I^* + CF_3I$	$\longrightarrow$	$CF_3I + I$	+	
4.	$I^{(*)} + I^{(*)} + CF_3I$	$\longrightarrow$	$CF_3I + I_2$	+	
5.	$I^{(*)} + I^{(*)} + I_2$	$\longrightarrow$	$2 I_2$	+	
6.	$I^{(*)} + I_2$	$\longrightarrow$	$I_2 + I$	+	(22)
7.	$I^* + O_2$	$\longrightarrow$	$O_2 + I$		
8.	$I + CF_3$	$\longrightarrow$	CF_3I	O	
9.	$CF_3 + CF_3$	$\longrightarrow$	C_2F_6	O	

O Fast processes, important during the photolysis and immediately thereafter
+ Reactions that are only important lateron

Reaction 8 which further increases the population inversion. Reactions 8 und 9 are competing for the CF_3 radicals. Reactions 1–7 reduce the I* concentration and therefore limit the storage of energy in excited iodine atoms. Among these quenching reactions, 1 is dominant as long as CF_3 radicals are present, while lateron quenching by CF_3I may predominate due to its large concentration. It has been stated frequently that laser action in this system is ultimately limited by the formation of molecular iodine which is a very efficient quencher for I* (Reaction 6). The I_2 formation with CF_3I, however, is relatively slow and inefficient and for the experimental conditions stated above has to be taken into account only at longer times. Once I_2 has been formed, processes 5 and 6 proceed with high efficiency. After completion of the fast Reactions 8 and 9, the gas mixture contains mainly CF_3I, C_2F_6, I* and I together with some impurities like O_2. From then on the further decay of I* is controlled by CF_3I deactivation. In addition the decay of I* will be very sensitive to the presence of I_2 and therefore to the rate of Reaction 4.

The discussion shows that one of the limitations to the storage of energy in the system is collisional deactivation. A parameter most sensitive to these processes is the optical gain of the amplifier system. Since the gain disappears as the population inversion $\Delta N = \left(N_{I*} - \dfrac{N_I}{2} \right)$ decays, gain measurements were chosen to investigate the chemical processes which determine the concentration of excited iodine atoms as a function of time. This technique of time-resolved gain spectroscopy [25,53] yielded data on the rate of energy-transfer processes involving excited iodine I*. These measurements will be discussed in Section 9 of this paper. A result worth mentioning here is that a plot of gain versus time definitely shows that the inversion continues to increase after termination of the photolysis flash. This effect, which has been reported before [51] and is probably due to additional chemical pumping processes, is not yet understood. It follows then that the set of reactions in Eq. (22) does not fully describe the chemistry of this laser system. In fact, the iodine laser may even be considered a chemical laser in part.

This laser shows some potential for high-power operation [54] which will be discussed in detail in Section 8 of this review.

6.2. Hydrogen Fluoride

Chemical laser action has so far been restricted to the molecules HF, HCl, HBr and their deuterated analogs and to CO. Lasing has also been stated in a brief report to occur in the OH radical produced in the O_3/H_2 photolysis [105]. The pumping scheme is likely to be

$$O_3 \xrightarrow{h\nu} O_2 + O(^1D) \qquad (23)$$
$$O(^1D) + H_2 \longrightarrow OH(v) + H$$

There has been no further information on this laser.

The hydrogen-fluoride laser, first described by Kompa and Pimentel [122] in 1967 and independently by T. Deutsch [123], has become the most popular chemical laser system. One might even say without exaggeration that HF now is the most carefully investigated molecule of all diatomics. The most frequently used reaction for pumping the laser is:

$$F + H_2 \underset{k'_v}{\overset{k_v}{\rightleftarrows}} HF(v) + H \qquad (24)$$

$$E_{tot} = -\Delta H + E_A + 5/2\,RT \approx 35\ kcal/mole$$

Other pumping steps are possible, for instance, in chain reactions and with other hydrogen- and fluorine-containing reaction partners. Extremely high gains have been found in this laser [124]. As outlined in Section 8, three types of processes have to be included for a full description of this laser: formation of the active HF molecules, relaxation and deexcitation reactions, and radiative processes. Each process has to be considered as function of the vibrational quantum number v and rotational quantum number J. However, even if only the v-dependence is included, the set of differential equations describing the temporal behavior of the system includes some sixty rate equations. All the rates in addition are more or less dependent on J. For obvious reasons, no account of the rotational effects has been published so far. In spite of all the rate information that has been accumulated, this aspect has not been explored sufficiently but may be important. The considerable complexity of this laser system calls for very extensive collaboration of theoreticians and experimentalists.

Table 5 lists most of the papers published on HF which contain experimental observations. Conference reports are quoted only where no other accounts have been published to date.

Table 5. Pulsed hydrogen-fluoride chemical lasers

$h\nu$ = flash photolysis
e = pulsed-discharge initiation

Reaction system	Type of information	Refs.
$UF_6 + H_2(D_2,\ HD)/h\nu$	First report of a flash initiated HF laser, empirical study of experimental parameters, quenching by various additives, spectra	Kompa, Pimentel Kompa, Parker, Pimentel [122]
$UF_6 + CH_4(CD_4,\ C_2H_6,\ C_3H_8,\ n{-}C_4H_{10},\ i{-}C_4H_{10},\ CH_3F,\ CH_2F_2,\ CHF_3,\ CH_3Cl,\ CH_2Cl_2,\ CHCl_3)\ /\ h\nu$	Comparison of hydrogen compounds, spectra	Parker, Pimentel [153]
$UF_6 + H_2(CH_4)\ /\ h\nu$	Vibrational energy partitioning study	Parker, Pimentel [112]

Table 5 (continued)

Reaction system	Type of information	Refs.
$UF_6 + CCl_3H$ / $h\nu$	Vibrational energy partitioning study	Parker, Pimentel [154]
UF_6 (XeF_4, SbF_5, WF_6) $+ H_2$ (CH_4) / $h\nu$	Comparison of fluorine sources, spectra	Kompa, Gensel, Wanner [155]
$WF_6 + H_2$ / $h\nu$	Comparison of different flash photolysis setups, actinometry	Gensel, Kompa, Wanner [28]
$WF_6 + H_2$ (D_2, CH_4, C_4H_{10}, HCl) / $h\nu$	Measurement of rate constants of fluorine atom reactions	Kompa, Wanner [136]
$IF_5 + H_2$ / $h\nu$	Indication of chain reaction with IF_5 as fluorine source, unknown transitions, spectra	Gensel, Kompa, Wanner [156]
MoF_6 (UF_6, F_2) $+ H_2$ / $h\nu$	Comparison of fluorine sources, experimental parameters	Dolgov-Savel'ev, Polyakov, Chumak [157]
$N_2F_4 + HCl$ (CH_4, CH_3F, CH_2F_2, CH_3Br, C_2H_6, C_2H_5F, C_2H_5I) / $h\nu$	Comparison of hydrogen compounds, analysis of pumping reactions	Brus, Lin [158]
$F_2O + H_2$ / $h\nu$	Use of F_2O as a fluorine source, possible chain reaction	Gross, Cohen, Jacobs [159]
$MoF_6 + H_2$ / $h\nu$	Pulse delay interpretation, MoF_6 actinometry, laser parameters	Chester, Hess [116]
$N_2F_4 + CD_4$ / $h\nu$	DF overtone emission	Suchard, Pimentel [74]
$CF_3I + C_2H_2$ (C_2H_4, C_2H_6, CH_4) / $h\nu$	Energy partitioning study	Berry [160]
ClF_3 (ClF) $+ H_2$ / $h\nu$	Investigation of chain reaction pumping	Krogh, Pimentel [161]
$WF_6 + CH_4$ / $h\nu$	Small-signal gain measurements	Gensel, Kompa, MacDonald [120]
$F_2O + H_2$ / shock wave	Reaction initiation by a shock wave	Gross, Giedt, Jacobs [162]

Table 5 (continued)

Reaction system	Type of information	Refs.
$ClN_3 + NF_3(SF_6) + H_2 / h\nu$	Thermal reaction initiation by ClN_3 explosion	Jensen, Rice [163]
$N_2F_4(NF_3) + H_2(B_2H_6) / h\nu, e$	High-energy electron-beam initiation, comparison with flash photolysis	Gregg et al. [164]
$CF_3I + CH_3I / h\nu$	Elimination laser produced by decomposition of chemically activated CF_3CH_3	Berry, Pimentel [165]
$H_2C = CHX / h\nu$ $(X = F, Cl)$	First demonstration of this type of molecular photoelimination laser	Berry, Pimentel [166]
$CH_3I + N_2F_4 / h\nu$	Elimination laser produced by decomposition of vibrationally excited CH_3NF_2	Padrick, Pimentel [139]
$C_2H_2O + CH_3F / h\nu$	Energy-partitioning study	Roebber, Pimentel [167]
$CH_2FCOCH_2F / h\nu$	Energy-partitioning study	Cuellar-Ferreira, Pimentel [168]
$O_3 + CH_nX_{4-n} / h\nu$ $(X = F, Cl, n = 1, 2, 3)$	Elimination laser action from the α-halo-methanols	Lin [169]
$CF_3I + HI / h\nu$	Energy-partitioning study	Coombe, Pimentel, Berry [192]
$H_2 + F_2$	Demonstration of purely chemical pumping	Spinnler, Kittle [170]
$H_2 + F_2 / h\nu$	Investigation of quantum yield	Burmasov, Dolgov-Savel'ev, Polyakov, Chumak [171]
$H_2 + F_2 / e$	Investigation of chain branching, efficiency estimates	Batovskii, Vasilev, Makarov, Talrose [172]
$H_2 + F_2 / h\nu$	Emission spectrum, "hot" versus "cold" reaction	Basov, Kulakov, Markin, Nikitin, Oraevskii [145]

Table 5 (continued)

Reaction system	Type of information	Refs.
$H_2 + F_2 \mid h\nu$ Laser	F_2 photolysis with second harmonic of ruby laser, quantum yield	Dolgov-Savel'ev, Zharov, Neganov, Chumak [70]
$H_2 + F_2 \mid h\nu$	Chain-reaction pumping	Hess [173]
$H_2 + F_2 + MoF_6 \mid h\nu$	Addition of MoF_6 to increase reaction rates	Hess [174]
$H_2 + F_2 \mid h\nu$	Time-resolved spectroscopy	Suchard, Gross, Whittier [132]
$H_2O + F_2 \mid e$	Demonstration of laser action with H_2O as the hydrogen source	Florin, Jensen [175]
$H_2 + F_2 \mid h\nu$	Atmospheric pressure operation by mixing above the second explosion limit	Wilson, Stephenson [176]
$CF_4(CBrF_3, CClF_3, CCl_2F_2)$ $+ H_2(D_2, CH_4, CH_3Cl) \mid e$ $Cl_2(Br_2) + H_2(D_2) \mid e$	First report of the discharge-initiated chemical HF laser, spectra, parameter study, rotational emission of HF, HCl	Deutsch [123,133,177]
$CHF_2Cl(CHFCl_2, CHF_3, CF_2Cl_2) + H_2 \mid e$	Demonstration of transverse multiple-arc discharge initiation	Lin, Green [178]
$NF_3 + H_2(C_2H_6) \mid e$	Demonstration of laser action	Lin [179]
$NF_3(N_2F_4) + H_2(CH_4, C_2H_6, HCl, HBr, natural gas)$ $\mid e$	Emission spectra, power measurements, investigation of population inversion in HF, R-branch lines	Lin, Green [180,181]
$SF_6(CCl_2F_2, CF_4) + H_2$ $(CH_4, C_3H_8, C_4H_{10}, C_6H_{14}, HI) \mid e$	New laser systems, optimization of experimental parameters, pulse energies	Jacobson, Kimbell [182]
$SF_6(C_3F_8, C_2F_6, CF_4)$ $+ C_3H_8(H_2, CH_4, C_2H_6, C_4H_{10}) \mid e$	Atmospheric pressure operation	Jacobson, Kimbell [183]

Table 5 (continued)

Reaction system	Type of information	Refs.
$SF_6 + H_2 / e$	Vibrational-rotational and purely rotational laser action of HF, comparison of 13 gases in the same experimental set-up, emission wavelengths 0.8—28 μ	Wood, Burkhard, Pollack, Bridges [184]
$SF_6 + H_2 / e$	Comparison of performance and spectra for HF, HCl, HBr and all isotopic compounds	Wood, Chang [185]
$SF_6 + H_2(HBr) / e$	Investigation of laser performance	Jensen, Rice [186]
$SF_6 + H_2 / e$	Parameter study, gain measurement, R-branch transitions in selective cavity	Marcus, Carbone [187]
$SF_6 + H_2 / e$	Design parameters, gain coefficients, spectra	Ultee [188]
$SF_6 + H_2(C_4H_{10}) / e$	Energy and efficiency has been investigated, double-discharge laser arrangement	Wenzel, Arnold [189]
$SF_6 + H_2 / e$	Superradiant emission, line-narrowing	Goldhar, Osgood, Javan [124]
$SF_6 + H_2 / e$	Spectroscopy, rotational energy distribution, energy > 1 J	Pummer, Kompa [71]
$SF_6 + H_2 / e$	Optical pumping of HF rotational lines	Skribanowitz, Osgood, Feld, Herman, Javan [73]
$NF_3 + SiH_4(GeH_4, AsH_3, B_2H_6, C_4H_8, C_8H_{10}, C_6H_{12}, C_3H_6, i-C_4H_{10}, n-C_4H_{10}, C_2H_6, CH_4, neo-C_5H_{12}) / e$	Comparison of hydrogen donors, parameter study	Pearson, Cowles, Herman, Gregg [190]
$BF_3(BCl_3, BBr_3) + H_2O / e$	Rotational emission in HF, HCl, HBr	Akitt, Yardley [191]

37

6.3. Other Hydrogen Halides

The hydrogen-halide lasers to be considered here besides HF are hydrogen chloride and hydrogen bromide and the corresponding deutero compounds. HCl lasers are normally generated according to the following scheme:

$$
\begin{aligned}
Cl_2 &\xrightarrow{\text{Diss.}} 2\,Cl \\
Cl + H_2 &\longrightarrow HCl + H & \Delta H &= +1\,\text{kcal} \\
H + Cl_2 &\longrightarrow HCl(v) + Cl & \Delta H &= -45\,\text{kcal}
\end{aligned} \tag{25}
$$

Further details of this reaction are discussed in Sextion 9. An important alternative reaction used for pumping is that of a Cl atom with HI.

$$
\begin{aligned}
Cl_2 &\xrightarrow{\text{Diss.}} 2\,Cl \\
Cl + HI &\longrightarrow HCl(v) + I & \Delta H &= -31{,}7\,\text{kcal}
\end{aligned} \tag{26}
$$

Hydrogen bromide formation may be described in an equally global fashion by

$$
\begin{aligned}
Br_2 &\xrightarrow{\text{Diss.}} 2\,Br \\
Br + H_2 &\longrightarrow HBr + H & \Delta H &= +16\,\text{kcal} \\
H + Br_2 &\longrightarrow HBr(v) + Br & \Delta H &= -41\,\text{kcal}
\end{aligned} \tag{27}
$$

Table 6 lists the known experimental results for these laser systems.

Table 6. HCl and HBr pulsed chemical lasers

Reaction system	Type of information	Refs.
$H_2 + Cl_2 / h\nu$	First report of a chem. laser, empirical investigation of emission details	Kasper, Pimentel [193]
$H_2(HD) + Cl_2 / h\nu$	Spectroscopy, investigation of pumping processes, discussion of detailes rate constants	Corneil, Pimentel [194]
1,1-(*cis* 1,2-) *trans* (1,2-) $C_2H_2Cl_2 / h\nu$	Energy-partitioning study	Berry, Pimentel [138]

Table 6 (continued)

Reaction system	Type of information	Refs.
1,1-(cis 1,2-) $C_2H_2Cl_2$ / $h\nu$	Energy-distribution study, tandem laser	Molina, Pimentel [113]
$H_2 + Cl_2$ / e $H_2 + Br_2$ / e $HI + Cl_2$ / e	Stimulated emission observed marginally in infrared luminescence study	Anlauf, Maylotte, Pacey, Polanyi [10]
$H_2 + Cl_2$ / e $H_2 + Br_2$ / e	First observation of pulse discharge-initiated HCl, HBr lasers, spectra	Deutsch [123,133,177,195]
$HI + Cl_2$ / $h\nu$	New reaction system with potentially very high inversion	Airey [196]
$HI + Cl_2$ / e	Initiation technique	Moore [197]
$Cl_2 + HBr$ / $h\nu$	Theoretical and experimental study, rate equations, computer modelling	Airey [99]
$Cl_2 + H_2$ / e	New reaction system, chain mechanism proposed	Lin [198]
$O_3 + CH_nX_{4-n}$ / $h\nu$ (X = F, Cl, n = 1, 2, 3)	Elimination laser action from the α-halo-methanols	Lin [169]
$Cl_2(NOCl) + H_2$ / $h\nu$	Small-signal gain measurements	Henry et al. [119]
$BCl_3(BBr_3, BF_3) + H_2O$ / e	Rotational emission in HCl, HBr, HF	Akitt, Yardley [191]
$H_2 + Cl_2$ / e $H_2 + Br_2$ / e	Transverse multiple-arc excitation, power, experimental characteristics	Burak, Noter, Ronn, Szoke [199]
$H_2 + Cl_2$ / e $H_2 + Br_2$ / e	Comparison of performance and spectra for (HF), HCl, HBr and all isotopic compounds	Wood, Chang [185]

6.4. Carbon Monoxide

The carbon-monoxide chemical laser exhibits a variety of pumping reactions. Laser action was first reported by Pollack in CS_2/O_2 photolysis [125]. The pumping scheme in this case is believed to be the following [126]:

$$CS_2 \xrightarrow{hv,\,e} CS + S \qquad \text{Initiation}$$
$$S + O_2 \longrightarrow SO + O \qquad \text{Propagation and}$$
$$O + CS \longrightarrow CO(v) + S \qquad \text{laser pumping}$$
$$\Delta H = -75 \text{ kcal/mole} \tag{28}$$

$$S + SO \longrightarrow S_2O$$
$$S_2O + O_2 \longrightarrow SO + SO_2$$
$$SO + SO \longrightarrow S_2O_2 \rightleftharpoons SO_2 + S \qquad \text{Termination}$$
$$S_2O_2 + SO \longrightarrow S_2O + SO_2$$

O_2 may be substituted by NO_2 with some minor changes [127]. Alternative pumping for CO sequences are [128,129]

$$O_3 \xrightarrow{hv,\,e} O_2 + O(^1D)$$

$$O(^1D) + C_3O_2 \longrightarrow 3\,CO(v) \tag{29}$$

or

$$O(^1D) + CN(X^2\,\Sigma^+, v') \longrightarrow CO(^1\Sigma^+ v) + N(^2D)$$

A very interesting way of pumping a CO laser has been opened up recently in the discharge-initiated combustion of acetylene [130] or cyanogen [131]. Table 7 lists the published work on CO chemical lasers, arranged according to the chemistry; it also reflects the history of the field.

6.5. Pumping by Energy Transfer

The CO_2 laser, first described by C. K. N. Patel in 1964 [65], involves as one possible pumping step the transfer of vibrational energy form molecular nitrogen to the asymmetric stretching vibration (001) of CO_2. The laser emission then occurs at either the (100) or the (020) state. The energy level diagram is seen in Fig. 16. Excitation is also possible by direct electron impact, or recombination and cascading to populate the (001) level. As is well known, the high cross-section for excitation, the low collisional deactivation rates, and the long radiative lifetime of the upper state of the laser transition are very favorable for high-power operation of this laser [66].

Table 7. Pulsed carbon-monoxide chemical lasers

Reaction system	Type of information	Refs.
$CS_2 + O_2 / h\nu$	First report of a chem. CO laser, spectroscopy, chemical pumping scheme	Pollack [200]
$CS_2 + O_2 / h\nu$	Analysis of pumping reactions, new lines	Gregg, Thomas [134]
$CS_2 + NO_2 / h\nu$	New reaction system	Rosenwaks, Yatsiv [127]
$CS_2 + O_2 / e$	Discharge initiation of the pumping sequence, spectra	Arnold, Kimbell [201]
$CS_2 + O_2 / e$	Pulse discharge initiated CO laser	Jacobson, Kimbell [202]
$CS_2 + O_2 / h\nu$	Chemistry and performance characteristics, computer simulation	Suart, Dawson, Kimbell [126]
$C_3O_2 + O_2 / e$	New reaction system, spectra, pumping mechanism	Lin, Bauer [203]
$O_3 + C_3O_2 / h\nu$	Pumping by the reaction of $O(^1D)$ with C_3O_2	Lin, Brus [128]
$O_3 + CN / h\nu$	Pumping by the reaction of $O(^1D)$ with CN	Brus, Lin [129]
$CS_2 + O_2 / h\nu$	Small-signal gain measurements, infrared chemiluminescence study	Hancock, Smith [204] Hancock, Morley, Smith [121]
$O_2 + C_2H_2 / e$	Discharge-initiated combustion of acetylene and other hydrocarbons	Barry, Boney, Brandelik [130]
$O_2 + C_2N_2 / e$	Generation of CO laser in cyanogen combustion with relatively high efficiency	Brandelik, Barry, Boney [131]
$CS_2 + O_2 / e$	Parameter study, spectroscopy, Q-switching	Ahlborn, Gensel, Kompa [205]

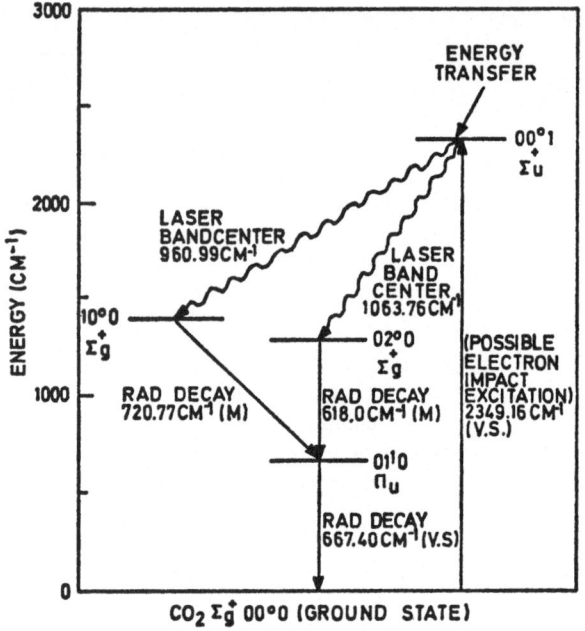

Pump transition		Coupled transition $(v = 1$ excited state)		
Designation[a]	Wavelength (μ)	J_{upper}	$-J_{\text{lower}}$	Wavelength (μ)
$P_1(3)$	2.608	2	1	126.5
$P_1(4)$	2.639	3	2	84.4
$P_1(5)$	2.672	4	3	63.4
$P_1(6)$	2.707	5	4	50.8
$P_1(7)$	2.744	6	5	42.4
$P_1(8)$	2.782	7	8	36.5

[a] $P_1(J)$ signifies the $(v = 1, J - 1)$ $(v = 0, J)$ transition Pumping scheme and wavelengths of far-infrared rotational oscillations for P-branch transitions in HF.

Fig. 16. Partial molecular-energy level diagram and transitions involved in the CO_2 laser (according to Patel[65])

Quite similar to this pumping scheme was an attempt to pump a CO_2 laser by chemical means reported by Russian workers [67]. Vibrationally excited N_2 was produced by flash-photolyzing hydrazoic acid (HN_3) and subsequent energy transfer. However, pumping by energy transfer from hydrogen halides was first achieved by R. W. F. Gross in 1967 [68]. The advantage of this type of laser resides in the principal shortcoming of all hydrogen halide lasers, namely the rapid vibrational relaxation. As a result, accumulation and storage of vibrational energy is very limited in HF or similar lasers. Transfer from the hot reaction product to the cold CO_2 admixture increases the efficiency and the output of pulsed as well as cw lasers. Such lasers have been most effectively produced in $D_2-F_2-CO_2$ reaction mixtures. The maximum pulse energies according to preliminary reports go up to 15 J. [69] The reactions of interest are

$$F + D_2 \longrightarrow DF(v) + D$$

$$D + F_2 \longrightarrow DF(v) + F \tag{30}$$

together with

$$DF(v) + CO_2 \longrightarrow CO_2(001) + DF(v-1) + \Delta E$$

For an analytical description of the laser, a detailed knowledge of the energy transfer rates is of key interest. Table 8 contains a summary of published transfer rates which are very large (corresponding to 50–200 collisions).

Table 8. Vibrational energy transfer in $HF-CO_2$ and $DF-CO_2$ mixtures

Process	k sec^{-1}torr^{-1}	Ref.
$DF(v = 1) + CO_2(000) \longrightarrow DF(v = 0) + CO_2(001)$	$17.75 \pm 2.5 \cdot 10^4$	[140]
	$58 \cdot 10^4$ [a]	[141]
$HF(v = 1) + CO_2(000) \longrightarrow HF(v = 0) + CO_2(001)$	$3.7 \pm 0.3 \cdot 10^4$	[140]
	$3.8 \cdot 10^4$ [a]	[141]
	$5.9 \pm 0.1 \cdot 10^4$	[143]
$CO_2(001) + DF(v = 0) \longrightarrow CO_2(mn0) + DF(v = 0)$	$1.9 \pm 0.4 \cdot 10^4$	[140]
	$1.86 \cdot 10^4$	[142]
$CO_2(001) + HF(v = 0) \longrightarrow CO_2(nm0) + HF(v = 0)$	$3.6 \pm 0.3 \cdot 10^4$	[140]
	$2.29 \cdot 10^4$	[142]
	$5.3 \pm 0.2 \cdot 10^4$	[143]

[a] Other values obtained with different assumptions.

Conventional CO_2 lasers are pumped in electric discharges. The method described here is different in that it uses flash photolysis for initiation of the pump sequence. An important feature is that one can utilize the potential of chain reactions in pumping such hybrid lasers.

6.6. Emission Spectra of Pulsed Chemical Lasers

There is a large number of chemical laser publications that deal with the investigation of the output spectra, mostly in a rather empirical fashion. The two questions which may be answered qualitatively concern the identification of the pumping reaction and the confirmation of predictions for the output composition. Such predictions may be based on spontaneous infrared luminescence measurements.

By investigating the vibrational transitions present in the laser signal, the contribution of the "hot reaction"

$$H + F_2 \longrightarrow HF(v) + F \qquad \Delta H = -98 \text{ kcal/mole} \qquad (31)$$

versus the "cold reaction"

$$F + H_2 \longrightarrow HF(v') + H \qquad \Delta H = -31 \text{ kcal/mole} \qquad (32)$$

in an HF laser may be determined. The terms "hot" and "cold" here refer to the fact that the first of these reaction steps, by virtue of its greater exothermicity, can pump higher vibrational states than the second one. The second reaction can populate only $HF(v \leqslant 3)$ states. As Table 9 shows, the contributions to the total output from higher vibrational transitions $(v > 3)$ are very small*. This was qualitively confirmed by another investigation yielding the vibrational distributions shown in Fig. 17 for HF and DF lasers. Since the inversions pumped by the hot reaction are expected to be considerable (compare Table 2), very fast vibrational relaxation of high v-levels has to be invoked to explain the very small high-v emission. Emission from $v = 5 \rightarrow 4$ and $v = 6 \rightarrow 5$ transitions has also been observed in the photolysis of IF_5/H_2 mixtures, giving indication of a chain reaction in this system, too.

Chemical-laser emission spectra up to 1967 have been compiled by Patel [65]. There is some inconsistency between HF laser spectra obtained in different laboratories and with different experimental set-ups. This is probably due in part to the absorption of several HF lines by the atmosphere inside or outside the laser cavity. However, there is additional inconsistency between the emission spectra of infrared chemiluminescence experiments and HF chemical lasers. While spontaneous luminescence predicts a peak in the $v = 2 \rightarrow 1$ transitions around $J = 6$ [14], the laser emission usually

---

* Note added in proof: More detailed investigations which have been conducted in the meantime indicate a stronger contribution from the hot reaction now (S. Suchard, private communication).

Table 9. Measured wavelengths, identification, and peak powers of HF laser transitions observed in flash photolysis of H_2 and F_2[132)]

Measured wavelength (μ)	Identification		
	Vibrational band	Transition (J)	Peak power (relative units)
2.61	$1 \to 0$	3	<1
2.64		4	<1
2.68		5	40
2.71		6	130
2.74		7	28
2.79		8	107
2.67	$2 \to 1$	1	6
2.70		2	300
2.73		3	600
2.76		4	300
2.80		5	62
2.83		6	8
2.87		7	<1
2.79	$3 \to 2$	1	165
2.82		2	170
2.86		3	417
2.89		4	379
2.93		5	4
2.96		6	14
3.01		7	2
3.05		8	8
2.92	$4 \to 3$	1	320
2.96		2	126
2.99		3	60

peaks at $J = 3,4$.[a)] This has been taken as an indication of excessive rotational relaxation prior to emission. The situation is different in HF lasers initiated in pulsed electric discharges [71)]. Emission spectra for different pressures are summarized in Table 10. The spectral output is very sensitive to pressure and gas composition. As the pressure is reduced, the total pulse duration is prolonged from 200 nsec to 35 µsec, and additional rotational lines which carry a considerable portion of the pulse energy then appear at high J numbers. Assuming rotational equilibrium, it is impossible to explain the appearance of these lines by a rise in the gas temperature. Thus it is concluded that the emission of this group of lines corresponds to a non-thermal rotational energy distribution.

---

a) There is one exception to this which has been reported recently [70)].

Table 10. HF laser spectra in SF_6/H_2 pulse discharge [71]

Transition		Pressures [Torr]		
$v \rightarrow v - 1$	P(J)	0.5 H₂ + 8 SF₆	0.25 H₂ + 3 SF₆	0.1 H₂ + 0.7 SF₆
		Energy [relative scale] [a]		
$1 \rightarrow 0$	P(3) [b]	—	—	—
	P(4) [b]	0.5	0.5	0.6
	P(5) [b]	0.25	0.3	—
	P(6) [b]	0.5	0.65	0.65
	P(7) [b]	1.6	1.45	1.8
	P(8) [b]	5.0	8.1	6.6
	P(9) [b]	4.5	4.15	6.8
	P(10)	—	—	—
	P(11) [b]	4.0	3.05	5.6
	P(12)	7.75	10.9	12.2
	P(13)	2.75	3.85	5.0
	P(14)	—	—	1.0
	P(15)	—	—	5.8
	P(16)	—	—	5.4
	P(17)	—	—	0.5
$2 \rightarrow 1$	P(3) [b]	0.5	0.3	0.6
	P(4) [b]	0.6	0.35	0.8
	P(5) [b]	1.0	0.65	1.0
	P(6) [b]	1.0	1.0	1.0
	P(7) [b]	3.3	2.9	6.0
	P(8) [b]	5.15	3.85	9.0
	P(9) [b]	3.5	1.1	7.0
	P(10)	—	—	—
	P(11)	2.35	2.4	6.0
	P(12) [b]	1,65	6.4	4.5
	P(13)	1.65	3.9	7.0
	P(14)	—	—	—
	P(15)	—	—	—
	P(16)	—	—	0.5
$3 \rightarrow 2$	P(2) [b]	0.1	0.3	0.4
	P(3) [b]	0.35	0.3	0.8
	P(4) [b]	0.45	0.3	0.65
	P(5) [b]	0.45	0.5	0.8
	P(6) [b]	0.45	0.85	1.2
	P(7) [b]	0.25	0.15	1.4
	P(8) [b]	0.45	0.65	0.6
	P(9)	0.1	—	—

[a] Normalized to $P_6 (2-1) = 1$ in each column. The ratio of the $P_6(2-1)$ lines for the for the three pressures in the above order is 1.7 : 1.2 : 1.

[b] Also observed at a pressure of 2.5 Torr H₂ and 60 Torr SF₆.

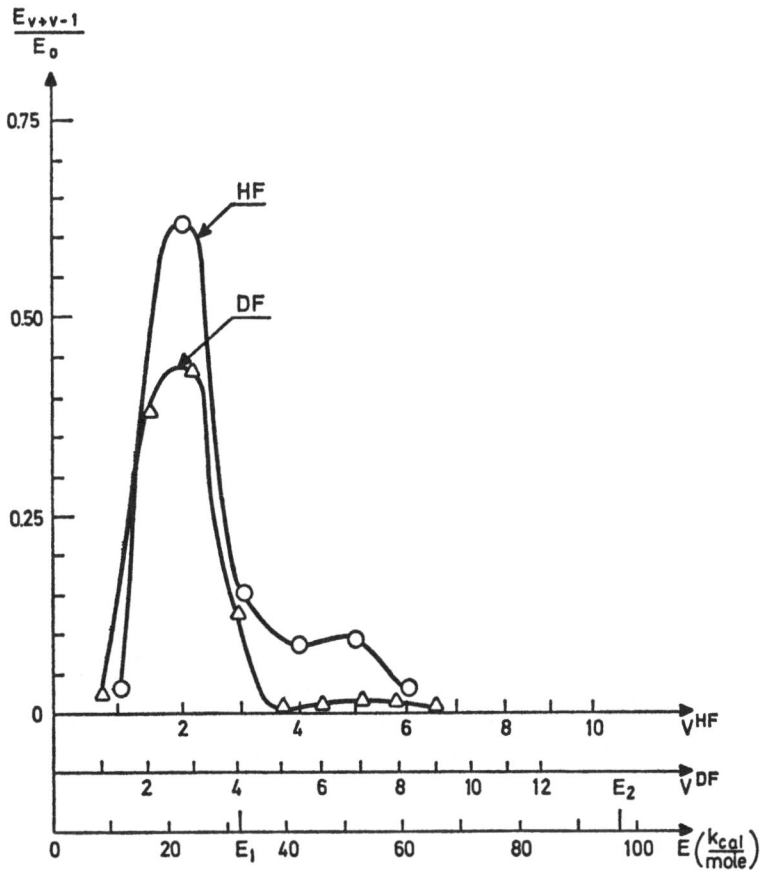

Fig. 17. Relative intensity contributions from the various vibrational transitions in HF and DF chain-reaction lasers. The major part of the emission corresponds to pumping by the "cold" reaction[145] (see text)

A tentative interpretation of these results may be based on the assumption that in this laser high rotational levels ($J > 10$) are indeed populated initially [71]. It should also be mentioned that two peaks in the rotational energy distribution are similarly observed in HCl chemiluminescence experiments under conditions of partial rotational relaxation [9]. A simple rotational relaxation model suggested by Polanyi and Woodall [72] has been applied to fit these observations. In this model the probability P of the relaxation process $(J + \Delta J) \rightarrow J$ is given as

$$P_J^{J+\Delta J} = C \exp -M (E_{J+\Delta J} - E_J) \,/\, kT \tag{33}$$

where C and M are constants.

Pure rotational HF laser emission can be produced under suitable conditions, as Table 11 shows. It is not clear at present, whether there

Table 11. HF rotational laser transitions observed using a pulsed discharge in a flowing freon and hydrogen mix ($CF_4 + H_2$ unless otherwise indicated) [133]

Measured wavelength (vacuum) μ	Identification		
	Vibrational level v	Lower rotational level J	
16.0215	0	15	(a)
14.4406	0	17	(b)
13.7841	0	18	
13.2009	0	19	
12.6781	0	20	
12.2082	0	21	
11.7854	0	22	
11.4033	0	23	
11.0573	0	24	
10.7439	0	25	
10.4578	0	26	
10.1978	0	27	
21.6986	1	11	(c)
20.1337	1	12	
18.8010	1	13	
15.0163	1	17	(b)
13.7277	1	19	
		or	
	2	20	
13.1877	1	20	
12.7006	1	21	
12.2619	1	22	
20.9393	2	12	(c)
14.2881	2	19	
13.2211	2	21	
10.8117	2	28	
10.5819	2	29	
21.7885	3	12	(c)
20.3513	3	13	
19.1129	3	14	
11.5408	3	27	
		or	
	4	29	

(a) $CCl_3F + H_2$
(b) $CClF_3 + H_2$
(c) $CBrF_3 + H_2$

the rotational emission depends to any extent on the vibrational rotational emission which occurs simultaneously. Rotational lines are also found in pulsed discharge-initiated HCl lasers.

HF rotational laser emission has also been obtained by Skribanovitz et al. [73] by pumping the first vibrational state of HF gas resonantly with another pulsed HF laser. Pumping the P-branch transitions of $v = 0 \rightarrow 1$ produces gain at the coupled rotational transitions in the $v = 1$ state. The principle is shown in Fig. 18. As stated in the above paper, the gain is very high and anisotropic and thus exhibits directional properties.

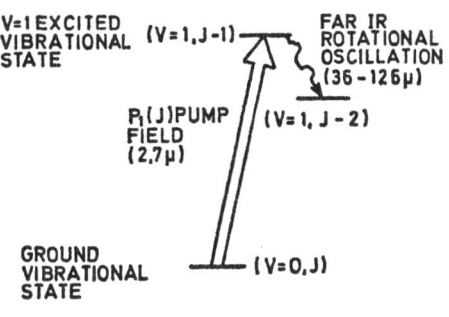

Fig. 18

	N₂	He
Area ratio (A/A^*)	15.3	15.3
Mach no. (M_j)	4.4	5.9
Pressure ratio (P_j/P_0)	0.0039	0.0018
Temperature ratio (T_j/T_0)	0.20	0.079
Velocity ratio u_j/α_0 (N₂)	1.99	4.80

Some effort has also been expended in analyzing the emission spectra of HCl, HBr (Table 6) and CO lasers. The CO spectra as produced in the CS_2/O_2 flash photolysis are given in Table 12. Measurements of the time of initiation of the lasing lines indicate that CO was being selectively excited at two vibrational levels by two different mechanisms, namely the chemical formation of excited CO and collisions with electronically excited SO_2.
It is noted that overtone emission ($\Delta v = 2$) has been achieved in a DF laser in $v = 3 \rightarrow 1$ transitions. [74]
In this context it also appears noteworthy that hyperfine splitting could be observed in the iodine photodissociation laser by Kasper et al. [146].

Table 12. Range of CO lines found to lase [134]

Vibrational transition	Lowest observed P-branch line	Highest observed P-branch line	Lowest observed R-branch line	Highest observed R-branch line
$1 \to 0$	9	21		
$2 \to 1$	6	20		
$3 \to 2$	8	33		
$4 \to 3$	8	22		
$5 \to 4$	8	33		
$6 \to 5$	8	33		
$7 \to 6$	7	32		
$8 \to 7$	7	33		
$9 \to 8$	6	33	6	8
$10 \to 9$	6	26	11	16
$11 \to 10$	7	27	13	22
$12 \to 11$	8	26	15	24
$13 \to 12$	9	28	2	12
$14 \to 13$	10	18	9	20
$15 \to 14$	12	16	15	28
$16 \to 15$	9	9		

6.7. Continuous Operation of Chemical Lasers

Earlier than with pulsed chemical lasers, the first technological break-through in chemical lasers occurred for continuous-wave lasers. Almost simultaneously in 1968 two groups successfully operated continuous-wave chemical lasers. One was at the Aerospace Corporation headed by T. A. Jacobs [75], the other one at Cornell University under T. A. Cool [76]. One of these lasers was an HF laser; the other was that is now called a hybrid chemical laser, being pumped by energy transfer rather than by a direct chemical reaction. This laser principle has been described in the context of pulsed chemical lasers in Section 6.5, In addition to these devices, an HF cw laser having millisecond flow duration was also demonstrated in principle in a shock tunnel. The latter employed diffusion of HCl into a supersonic stream containing F atoms [77].

The first cw HF lasers were operated in supersonic flows. This is deemed necessary for higher mass transport, smaller back diffusion from the reaction zone, and reduced collisional deactivation prior to emission. We will omit a description of the early stages of the development and present here some

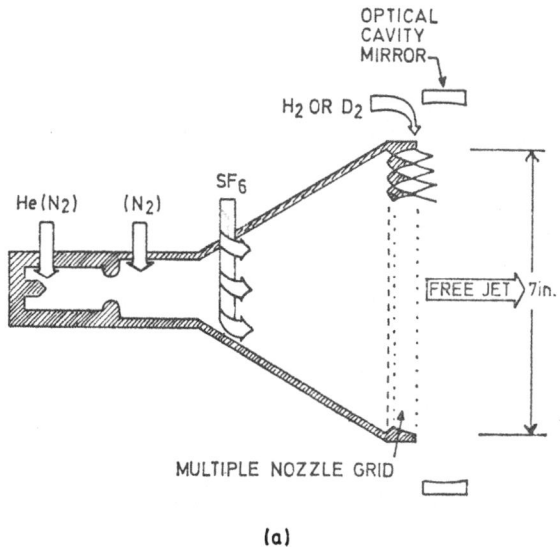

OPTICAL
CAVITY
MIRROR

H_2 OR D_2

He(N_2)   (N_2) SF_6

FREE JET 7in.

MULTIPLE NOZZLE GRID

(a)

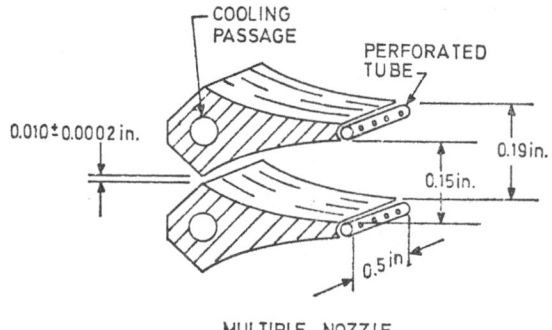

COOLING
PASSAGE

PERFORATED
TUBE

0.010 ± 0.0002 in.

0.19in.

0.15in.

0.5 in.

MULTIPLE NOZZLE

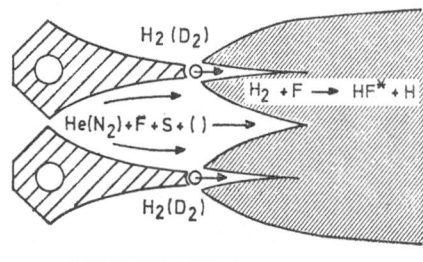

H_2 (D_2)

$H_2 + F \longrightarrow HF^* + H$

He(N_2)$+F+S+$( )

H_2(D_2)

DIFFUSION MIXING

(b)

Fig. 19. Schematic representation of an HF (DF) laser with supersonic flow (a). In part (b) nozzle and diffusion details are shown

operational characteristics of a high-power HF laser reported by Mirels, Spencer *et al.* [78]. In their system, arc-heated N_2 is mixed in a plenum with SF_6 to provide F atoms by thermal dissociation. The mixture is expanded to form a supersonic jet into which H_2 is diffused. Population inversion and lasing are due to

$$H_2 + F \longrightarrow HF(v) + H, v \leqslant 3, \Delta H = -31.7 \text{ kcal.}$$

Power levels above 1 kW are reported. The efficiency of emission of chemical energy to laser power is 16% at low SF_6 flow rates and approximately 10% at peak power. Fig. 19 gives a schematic representation of some of the operational features. It is intuitively obvious that, in order to have an efficient laser, it is necessary that the rate of H_2 diffusion into the jet and the rate of the pumping reaction be rapid compared with the rates of collisional deactivation. The performance of a corresponding DF laser has also been investigated [78]. The ratio of DF to HF laser power is 0.7 under similar flow conditions. The observed output spectra are reproduced in Table 13. It has been suggested that the lower DF efficiency is due to vibrational deactivation by N_2. The efficiency and intracavity power of HF and DF is indeed the same with He as a diluent instead of N_2. The efficiency of HF lasers with He and, with N_2 carrier gases is compared in Fig. 20.

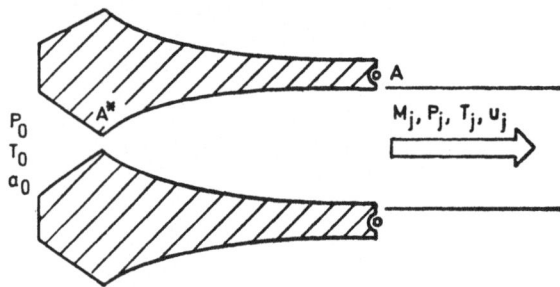

Fig. 20. Comparison of N_2 and He jet flow parameters

Obviously SF_6 as a fluorine source can be substituted by molecular fluorine. An improved version of this laser would thus employ partly dissociated $F_2 (T \sim 1000 °K)$ and a diluent in the plenum section. The F_2 present in the jet will permit the reaction $H + F_2 \rightarrow HF(v) + F, v > 3$, $\Delta H = -98$ kcal. The chemical potential of the $H_2 + F_2$ chain reaction

Table 13. Observed spectra from continuous-wave chemical lasers [135]

HF			DF		
Identification vibrational band[a]	Line	Wavelength (μ)	Identification vibrational band	Line	Wavelength (μ)
$1 \to 0$	P(4)	2.640	$1 \to 0$	P(8)	3.680
	P(5)	2.673		P(9)	3.716
	P(6)	2.707		P(10)	3.752
	P(7)	2.744		P(11)	3.790
				P(12)	3.830
$2 \to 1$	P(4)	2.760	$2 \to 1$	P(8)	3.800
	P(5)	2.795		P(9)	3.838
	P(6)	2.832		P(10)	3.876
	P(7)	2.871		P(11)	3.916
				P(12)	3.957
			$3 \to 2$	P(8)	3.927
				P(9)	3.965
				P(10)	4.005
				P(11)	4.046

[a] In later work also $v = 3 - 2$ emission of HF was reported [78].

could be realized in this way. The partial dissociation of F_2 in the plenum can be effected by combustion, regenerative heating, arc heating, etc. Experiments with such systems have been reported by Meinzer et al. [79]

Hydrogen halide cw lasers can also be operated in subsonic flows [80,81]. This means some sacrifice in power and efficiency but may be more convenient experimentally for some applications. Such lasers provide multiline power outputs ranging from 0.1 to 5 W for small devices requiring pumps of moderate size. Selective single-wavelength operation is possible with dispersive resonators. For pumping these lasers several simple atom-exchange reactions of the type extensively studied by Polanyi and coworkers have been employed. Laser radiation was obtained from HF, DF, HCl, and CO. Presumably HBr, DBr, DCl, and OH will be added to this list in the near future. Some of the results according to Cool et al. are summarized in Table 14. In all cases only partial population inversions have been found under these conditions. The experimental arrangements for these lasers are not different in principle from those used for chemically pumped CO_2 lasers.

An interesting laser device to be mentioned here is the CO laser resulting from a free-burning additive CS_2/O_2 flame. With the burner injector having an area of 30.5 by 0.79 cm, a maximum power of 0.6 W was obtained [82].

Table 14. Hydrogen-halide cw laser emission in sonic flows [80]

Reaction	Identification			Vibrational population ratio	Effective vibrational temp. (°K)
	Vib. band	Transition	Strength		
F + H$_2$	2 → 1	P(6)	w		
		P(5)	s		
				0.656	13 000
		P(4)	s		
		P(3)	w		
H + F$_2$	2 → 1	P(6)	w		
		P(5)	s	0.656	13 000
		P(4)	w		
F + Hl	2 → 1	P(5)	s		
	3 → 2	P(5)	w		
		P(4)	s		
F + D$_2$	3 → 2	P(8)	w		
		P(7)	s		
				0.685	10 300
		P(6)	s		
		P(5)	w		
	2 → 1	P(8)	s		
D + F$_2$	3 → 2	P(8)	w		
		P(7)	s		
				0.685	10 300
		P(6)	s		
		P(5)	w		
F + Dl	4 → 3	P(5)	s		
	3 → 2	P(6)	w		
H + Cl$_2$	1 → 0	P(6)	w		
		P(5)	s		
		P(4)	w		
Cl + Hl	2 → 1	P(7)	w		
		P(6)	s		
				0.770	17 000
		P(5)	s		
		P(4)	w		
	1 → 0	P(7)	w		
		P(6)	s		
				0.778	17 400
		P(5)	s		
		P(4)	w		

Efficient, purely chemical laser operation is possible in hydrogen halide-CO_2 transfer lasers, as developed by Cool and coworkers [83]. In these lasers no external energy sources are required. The systems, which operate on the mixing of commercially available bottled gases, are $HCl-CO_2$, $HBr-CO_2$, $DF-CO_2$, and $HF-CO_2$. The pumping scheme for a $DF-CO_2$ laser, for instance, is as follows:

$$F_2 + NO \longrightarrow NOF + F$$
$$F + D_2 \longrightarrow DF(v) + D \qquad (34)$$
$$DF(v) + CO_2 \longrightarrow CO_2(001) + DF(v-1)$$

Details of the energy transfer from DF to CO_2 have been discussed. The experimental set-up is schematically shown in Fig. 21. The power range reported so far is a few hundred watts for sonic flows. Extensions to achieve multikilowatt $DF-CO_2$ laser operation in supersonic flows with full atmospheric pressure recovery in the exhaust gases have been anounced recently [84].

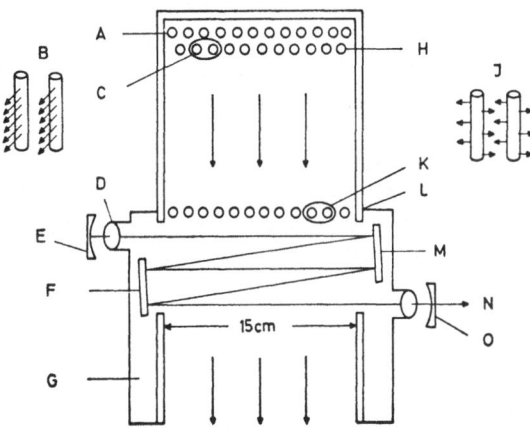

Fig. 21. Schematic representation of a subsonic CO_2 laser with purely chemical excitation (after Cool[82]). *A* He and F_2 injectors, *H* CO_2 and NO inlet, *C* construction detail shown in *B*, *L* D_2 mixing array, *K* part of the D_2 inlet system which is shown in detail in *J*, *D* sodium chloride window, *E* totally reflecting cavity mirror with long focal length, *M*, *F* beam-folding (plane) mirrors, *O* partially reflecting cavity mirror for output coupling, *N* laser beam, *G* resonator housing flushed with nitrogen

7. Future Chemical Lasers

Various directions may be distinguished in chemical laser research. There are some relatively straightforward attempts to utilize the potential of chain reactions, especially branched chains. There is a widespread and mostly empirical search for new laser reactions and new types of chemical lasers. Finally, there is a more fundamental approach to the problem on the basis of theories about the creation of inversions in reactive processes. Progress in this field will also depend largely on a better understanding of energy-transfer processes. Chemical laser research has already motivated many energy-transfer studies and has supplied a strong stimulus in this area (compare Section 3).

For the exploitation of chain reactions, we follow the discussion of Basov et al. [1]. With reference to the discussion of population inversions in Section 4 (14), we consider first the relationship between chemical pumping $P(t)$ and relaxation $L(t)$ which determines whether the reaction proceeds with or without an inversion. The temporal dependence of the level populations is given by the set of balance equations (16). Various types of temporal dependence of the pumping function $P(t)$ now have to be investigated. If the initial external energy input produces a certain concentration of active centers n, the pumping rate in the case of a linear chain reaction is given as

$$P(t) = k \, A \, n \, \exp - w_- t , \qquad (35)$$

where k is the rate of chain propagation, A is reagent concentration, and w_- is the chain termination rate. There are now two critical cases, characterized by $w_- < L(t)$, $w_- > L(t)$. It can be shown that in the first case the maximum population inversion ΔN is given by

$$\Delta N = (1 - \ln 2) \frac{kA}{L(t)} n \qquad (36)$$

while in the second case

$$\Delta N = \frac{kA}{w_-} n \qquad (37)$$

56

The quantity

$$\nu_{\text{opt}} = (1 - \ln 2)\frac{kA}{L(t)} \tag{38}$$

in the first case and

$$\nu_{\text{opt}} = \frac{kA}{w_-} \tag{39}$$

in the second case determines the number of molecules which contribute to radiation at each active center and may be called the optical length of the chain. These equations indicate that the inversion maximum depends on the rate of chain development. A substantial inversion can be achieved only if the latter exceeds the rate of vibrational relaxation. If $\nu_{\text{opt}} < 1$ in the case of strong relaxation, the chain effect remains unused and a large energy input is necessary to a sizable inversion density.

The rate of chemical pumping in the case of branched chains can be written as

$$P(t) = P_0 \exp st, \tag{40}$$

where s is the branching factor. This expression shows that the population inversion can grow exponentially with time if $s > L(t)$. The inequality $s > 0$ also defines the self-ignition condition. In the case where the inversion and ignition regimes are non-coincident a certain amount of energy must be supplied in initiating the reaction in order to bring it into the inversion region. In principle the reaction mixture could heat itself into and also out of the inversion region. However, if the inversion and ignition regimes coincide, an ideal chemical laser might be realized which could function with very little external energy. There are thus sound reasons for examining the extent to which branched-chain reactions can be exploited in chemical lasers.

A particularly interesting example here is the HF laser from H_2/F_2 mixtures where it is uncertain how much chain-branching occurs. The contributions of the chain steps could de betermined by an investigation of the laser emission spectra, as discussed in Section 6.6 of this review. A measurement of the quantum yield of an HF laser was conducted by Dolgov-Savel'ev et al. in experiments with the second harmonic of the ruby laser [70]. This has advantages over measurements with flashlamp initiation where a number of difficulties arise due to problems in determining the amount of energy absorbed in the mixture. The absorption coefficient of F_2 at the doubled ruby wavelength $\lambda = 3470$ Å was measured. For an absorbed energy of 5 mJ, the stimulated emission energy was found to be 100 mJ. Therefore $E_{\text{emission}}/E_{\text{absorption}} = 20$ and the quantum yield of generation was

$$E_{\text{emission}}\lambda_{\text{emission}}/E_{\text{absorption}}\lambda_{\text{absorption}} = 180 \,.$$

The quantum yield obtained in this manner depends on the experimental conditions. It is still not clear how much chain-branching contributes to the observed chain length of > 180 steps. Since chain-branching in the H_2/F_2 system relies on either energy transfer or thermal branching, attempts have also been made to explore the possibility of material chains by substituting for H_2 hydrocarbon components in which the generation of more than one radical per reaction step can produce branching. An example is the reaction with CH_2F_2 according to the reaction scheme

$$CH_2F_2 + F \longrightarrow CHF_2 + HF$$
$$CHF_2 + F_2 \longrightarrow CHF_3 + F - \Delta H > 80 \, kcal$$
$$CHF_3 \longrightarrow CH_2 + HF \qquad (41)$$
$$CF + F_2 \longrightarrow CF_3 + F$$
$$CF_3 + F_2 \longrightarrow CF_4 + F .$$

The chemical laser systems suggested by this reaction and other reactions of the same type are numerous, and more work along these lines is to be expected.

The list of unsuccessful attempts to find new chemical laser reactions is very long and will not be discussed in detail here. The reader is referred to the discussion of prospects at the first chemical laser conference which appeared as a supplement to Applied Optics [32]. A new approach of a more general nature is the photorecombination laser first suggested by R. A. Young [85] as early as 1964 and treated in detail by Kochelap and Pekar [86]. In describing this principle, we follow in part the argument of A. N. Oraevskii [87]. A number of chemical processes give rise to the emission of a photon such that this emission is not a consequence but a necessary condition of the elementary act.

$$A + B \;\rightleftharpoons\; (AB^*) \longrightarrow AB + h\nu \qquad (42)$$

(42) is an example of a photorecombination or photoaddition reaction since the stabilization of AB^* is possible only if the excess energy is given off as emitted radiation. The argument has been extended by Pekar to include photosubstitution reactions, but the discussion here is restricted to the photorecombination case. Gas-phase free-radical association almost inevitably populates highly excited states of the associative complex. The emission spectrum produced by the radiative recombination is usually continuous and does not give a high emission cross-section for any given frequency interval. Let us take a specific example to illustrate the difficulties: Let τ_d be the lifetime of the complexes AB^* with respect to dissociation

into A and B, and $1/\tau_{rad}$ the probability of emission by the complex. The rate of change of the density of AB^* is then

$$\frac{d\,AB}{dt} = k\,A\,B - \frac{1}{\tau_d}\,AB^* - \frac{1}{\tau_{rad}}\,AB^*$$

$$\frac{d\,A\,(B)}{dt} = -k\,A\,B + \frac{1}{\tau_d}\,AB^*,$$

(43)

where k is the collision rate constant. For a collision pair AB^* whose lifetime τ_d is only of the order of one vibrational period $\tau_d \sim 10^{-13}$ sec, and with an emission probability $W \sim 10^6$ sec^{-1}, one finds for an initial density of $A = B = n_0 = 10^{19}$ cm^{-3} that about half of the molecules have reacted in a time of $\sim 10^{-2}$ sec. Although this does not look encouraging, improvements are possible in two ways. First, one might find complexes AB^* which are not just collision pairs but transient species with a lifetime in the nsec or μsec range.

Alternatively the radiative lifetime τ_{rad} may be decreased by stimulated emission to

$$\frac{1}{\tau_{rad}} = \frac{1}{\tau_{rad(0)}}[1 + (\lambda^3 I/hc)]$$

(44)

where λ is the wavelength of radiation and I is the effective light intensity. I can be increased deliberately if an intense stimulating radiation field can be provided, for instance by an external laser source. This would suggest that such a photorecombination system should be operated as an amplifier rather than as an oscillator in order to demonstrate optical gain. In such a laser the increasing radiation intensity speeds up not only the stimulated transitions but also the reaction itself.

So far, preassociation and the possibility of photorecombinative laser action have been investigated in the formation of NO [85,88)], N$_2$ [85)], CN [85)], and the halogens [87)]. A somewhat different system is the "dimol" emission from an excited-state dimer of molecular oxygen in either the $^1\Sigma_g^+$ or $^1\Delta_g$ state [89)].

$$O_2(^1\Delta_g) \longrightarrow O_2(^3\Sigma_g^-) + h\nu_{12686}\,\text{Å}$$

$$\frac{1}{\tau} = 2.58 \times 10^{-4}\ \text{sec}^{-1}$$

(45)

$$2O_2(^1\Delta_g) \rightleftharpoons (O_2)_2^* \longrightarrow 2O_2(^3\Sigma_g^-) + h\nu_{6340}\,\text{Å}$$

$$\frac{1}{\tau} = 0.67\ \text{sec}^{-1}$$

The transition probability is seen to be considerably increased in the excited complex. In addition, the terminal state of the transition decomposes rapidly into two ground-state O_2 molecules. Finally, it should be mentioned that sizable densities of $O_2(^1\Delta_g)$ can be generated by various means, for instance, by a microwave discharge in O_2, the photolysis of ozone [90], or the alkaline halogenation of hydrogen peroxide [91]. More detailed consideration shows that successful laser operation will strongly depend on the lifetime τ_d of the complex $(O_2)_2^*$ which does not seem to be known with certainty. Laser action would appear to be feasible only if this complex could be accumulated to some extent.

These considerations lead to a general discussion of exciplex lasers. This is a well-established laser scheme for dye lasers where charge-taransfer complexes of excited molecules form the upper lasing state. Examples are the dyes 4-methyl-umbelliferone or N-methylacridin [92].

Such exciplex laser systems could perhaps also be pumped by chemical reactions in systems of the type investigated by Weller et al. [93].

$$\tag{46}$$

Our last example of this class of lasers is the xenon laser recently reported by Basov et al. [94]. Here stimulated emission in the far-UV region of the spectrum (1680—1700 Å) is generated by excitation of liquid Xe with high-energy electrons. The upper level of the transition is a bound excited level of Xe_2 that radiates to the repulsive ground state. The electrons used for the pumping first produce Xe*, Xe$^+$ and secondary electrons [95]. By collisional processes the energy is channelled down to the relatively stable excited molecular level where it piles up. The system is noteworthy, despite some difficulties in operating it, not only for its extreme wavelength but also for its potentially high efficiency (65%). The exploration of other exciplexes made up of molecules or atoms is suggested by these results. These examples show how the laser principle can be approached from rather different areas of chemistry.

Turning away from these somewhat exotic systems, we conclude with some brief remarks on the systematic theoretical study of the generation of inversions in the course of chemical reactions. Vibrational inversions are mainly concerned. J. C. Polanyi has discussed the connection between the properties of potential surfaces of exchange reactions and the formation of vibrational excitation and vibrational inversion [96]. In the normal case

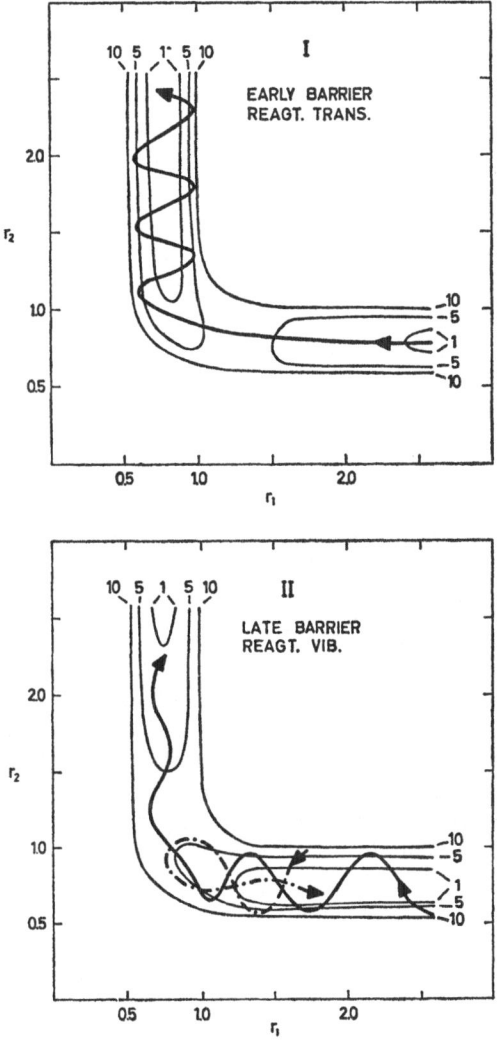

Fig. 22. Computed trajectories showing the role of barrier location in the formation of vibrationally excited products (after Polanyi[96]). The barrier heights on both surfaces are equal. In the upper case the reagent vibrational energy was zero while in the lower case a certain amount of reagent vibrational and translational energy was assumed. The unreactive trajectory had the same energy but the opposite vibrational phase. It is seen that a barrier along the approach coordinate is best surmounted by motion along that coordinate (reagent translation). A barrier along the separation coordinate is best surmounted by motion along that coordinate, which is to say, by reagent vibration

of an exchange reaction $A + BC \rightarrow AB(v) + C$ shown in Fig. 22, the factor determining the energy distribution among the reaction products appears to be the location of the energy barrier along the reaction coordinate. If this is located early in the entry valley, vibrational excitation of the product AB will result. There are exceptions to this rule and a much more detailed discussion is needed to cover all the possible cases. It can be concluded that in the situation shown in Fig. 22 translational energy in the reagents will be most effective in pushing the system over the barrier. Employing the argument of microscopic reversibility, it may then be said that in the reverse reaction $AB + C \rightarrow BC + A$ vibrational excitation will be most influential in bringing about the reaction. The opposite will be true of reactions in which the barrier is located late in the exit valley. Thus this model not only explains qualitatively the formation of inversions but also gives some indication of the relative effectiveness of translational and vibrational excitation in promoting the reaction.

A different approach to the problem of vibrational inversion in exchange reactions is found in the work of Hofacker and Levine [97]. These authors have shown that coupling between vibrational and translational degrees of freedom can be understood mainly in terms of the internal centrifugal force due to the curvature of the reaction path on the potential energy surface and the kinetic energy of the system. Conditions for a collinear collision complex, which is most favorable for inversion among products, and predictions of the model for enhanced inversion are: few atoms (effectively 3) in the collision complex, small reaction cross-section, light central atom in the complex potential energy surface with early attracting entry valley, exothermic reaction. In general, collision complexes with a minimum number of coupled degrees of freedom are the most promising producers of vibrational inversions. The concept of internal centrifugal force can be tested and may be very useful in selecting reactions with optimum potential for chemical-laser action. Some support for this theory is obtained from experiments on energy partitioning in the reactions $Ba + O_2$ and $Sr + O_2$ measured in a molecular beam experiment [98].

8. Present Perspectives of High-Power Chemical Lasers

The question of the high-power potential of chemical lasers has a straight-forward answer for cw lasers. Such lasers can operate in the kW range, as outlined in the preceding chapters. Where pulsed chemical lasers are concerned, there are striking differences between the expectation based on theoretical models and the actual performance of such devices. Many computer simulation studies have been published since Airey's first attempt to model an HCl laser [99]. We will follow here the discussion of Kerber, Emanuel and Whittier [108], again taking the HF laser as the model system. The processes which describe the growth and decay of the population inversion are the following:

a) H_2-F_2 chain

$$F + H_2 \rightleftharpoons HF(v) + H \tag{47}$$

$$H + F_2 \rightleftharpoons HF(v) + F \tag{48}$$

b) Vibrational-translational (VT) deactivation

$$HF(v) + M \rightleftharpoons HF(v-1) + M \tag{49}$$

c) Vibrational-vibrational (VV) quantum exchange

$$HF(v) + HF(v') \rightleftharpoons HF(v+1) + HF(v'-1) \tag{50}$$

$$HF(v) + H_2(v') \rightleftharpoons HF(v+1) + H_2(v'-1) \tag{51}$$

d) Dissociation-recombination

$$F_2 + M \rightleftharpoons M + F + F \tag{52}$$

$$H_2 + M \rightleftharpoons M + H + H \tag{53}$$

$$HF(v) + M \rightleftharpoons M + H + F \tag{54}$$

Most of the rate constants for the processes listed here are known from experimental studies or can be estimated with reasonable accuracy. With certain assumptions, a detailed kinetic model can be developed including rate equations for the growth and decay of the HF vibrational-state densities. To consider the stimulated processes, a Boltzmann distribution is assumed for the rotational levels with lasing on each vibrational band at the line center of the transition having maximum gain. Given these assumptions, only one $v-J$-transition within a band can lase at a given time. Initiation of the reaction is simulated by the instantaneous introduction of some F atom concentration. Excited species are then produced according to the kinetic parameters until, at some time and for some J, the gain on a vibration-rotation transition reaches threshold. From then on the gain for the highest gain transition is equal to the threshold gain $a_0 = \sigma \Delta n_0$ (compare (17), (18)) which is given as

$$\alpha_0 = - \ (1/2 \ L) \ln (R_1 \ R_2)$$

ignoring transmission losses. The somewhat questionable assumption of a uniform photon density in the laser cavity is made, and the gain is assumed to be constant over the amplifier length.

The interaction of the cavity and chemical mechanisms may be predicted by means of such calculations. Since lasing has a large effect on the concentration of excited HF. The importance of deactivation mechanisms differs for the zero power case and during lasing.

Of the 31.56 kcal/mole released by Reaction (47), 22.6 kcal is channeled into vibrational excitation of HF (v) (compare Table 2). Reaction (48) in addition produces 98.04 kcal/mole of which 43.1 kcal is available as vibrational energy. Thus 50.7% of the chemical energy of the chain is converted to vibrational energy. Obviously not all of this energy could be extracted as laser radiation. However, in the computations considered here an efficiency of $\sim$20% was found to be possible.

It is somewhat difficult to compare these predictions with experimental results since no really systematic experimental study has yet been published. This is due in part to difficulties in preparing mixtures of H_2 and F_2 of any desired composition and pressure and also to experimental limitations in the sufficiently rapid initiation of the pumping reaction. However, as far as the experimental information goes, it can be concluded that the efficiency is considerably lower than expected. For instance, in flash photolysis-initiated HF lasers a chemical efficiency of below 1% is usually found [101]. Two suggestions may be made to explain this discrepancy. One may look at it as either a chemical rate problem or a laser problem. In the first case, some unknown rate process must be assumed to reduce the build-up of excited HF. Since the formation and deactivation rates are known with some accuracy, this could only be excessive recombination or an unusually high rate of the reverse reaction [102]. Alternatively, parasitic oscillations or superradiance have been claimed to cause radiation losses in off-axis directions. The reason for this may be seen in the potentially very high gain of HF lasers. Since the stimulated emission cross-section σ, which has not been measured experimentally, might attain values as high as 10^{-16} cm^2, it follows that the small signal gain $V = \exp \sigma \Delta n l$ can become very high even with relatively low inversion densities Δn. Thus parasitic modes bouncing off the laser tube walls which do not experience much feedback from the resonator mirrors could be preferred due to their longer amplifying length. In the most extreme case this would mean that there is no longer any defined cavity mode structure because induced emission can occur in any direction.

This interpretation is supported by the observation that at low pressures relatively good agreement is found between experiment and theory. As the pressure is raised, an abrupt levelling-off of peak intensity occurs whereas the calculations predict a continuing increase [103].

It appears that we need a more detailed insight, both experimental and theoretical, into the collisional and radiational rate processes. A saturated amplifier experiment should be done in order to monitor the total amount of inversion in the chemical reaction system. In the discussion of the balance Eq. (16) in Section 4 of this review, it was stated that in principle it is always possible for the stimulated processes to predominate over the collisionally controlled reactions. In such a large-signal gain measurement all the inversion that is created immediately contributes to the amplification since a sufficiently intense stimulating radiation field gives rise to a high stimulated-emission rate. Unlike the small-signal gain region referred to above, the large-signal amplification V is given as

$$V = E_l / E_0 = 1 + \frac{\Delta n \, h\nu}{\left(1+\frac{g_1}{g_2}\right) E_0} \tag{55}$$

E_0 is the input energy, E_l the output energy of the amplifier, g_1, g_2 are the degeneracies of the states. The amplification then depends only on the total inversion ΔN [cm^{-2}] (compare Section 9.3). The experiment suggested here aims to reduce or eliminate any collisional quenching and control the emission by saturation of the amplifier medium at all times.

The question concerning chemical efficiency is even harder to answer for chemical lasers initiated by electric discharge. Here it is uncertain how much of the electrical energy is effectively deposited in the medium. The maximum electrical efficiency that has been reported is $\sim 4\%$ in discharge lasers operating on SF$_6$/H$_2$ mixtures [104]. Pulse energies of > 10 J have been reported with a double-discharge technique and 10 J pulses were obtained in the author's laboratory with a simple multiple-pin TEA type laser. Obviously lasers with such pulse energies and suitable repetition rates are useful tools for many applications. They suffer, however, from the same problems as TEA CO_2 lasers. It is difficult to excite larger volumes of gas without arc formation. Due to the highly electronegative character of SF_6, high-pressure operation is even more difficult. These problems have been overcome in CO_2 lasers by using electron-beam controlled discharges. Accordingly there are tendencies in several laboratories to use such sustained discharges in HF lasers, too. Evidently the next steps in this development could be the exploration of fluorides other than SF_6 and perhaps of chain reactions.

A principal shortcoming of all hydrogen-halide lasers is their rapid vibrational deactivation. The deactivation problem is much less severe if the vibrational energy of, for instance, deuterium fluoride is transferred to CO_2 as an admixture. A chemically pumped CO_2 laser is operated in this fashion. The principle of operation has been considerably advanced by Cool and coworkers for cw lasers, as discussed above (Section 6.4). Recently this type of transfer laser has yielded energies of 3 J and 5—15 J in pulsed lasers, too. Considerably higher energies are to be expected from this type of laser. One might envisage the combination of an electrically driven CO_2 laser oscillator with a chemically pumped CO_2 laser amplifier.

For many applications of high-power lasers not only are large energies important but it is also essential to have a short pulse duration. This can be accomplished to some extent by very rapid pumping in combination with pulse-shortening in an amplifier chain [106]. For instance, 20 nsec pulses have been obtained in HF lasers in a fast discharge. Pulses in the nsec and sub-nsec range, however, will usually require some sort of optical manipulation or switching technique. In this case the energy has to be stored for some time before being released. There are two main prerequisites for this: collisional deactivation mechanisms must not be too effective so that accumulation and storage of the energy is possible over a certain time range; and self-starting of the oscillation and/or excessive superradiance losses must be avoided. We will ignore here the material problems often encountered in high-power lasers.

Owing to their very rapid vibrational deactivation, hydrogen-halide lasers are not very suitable candidates for giant-pulse operation. The only chemical laser so far known which can meet these requirements is the photochemical iodine laser. A high-power iodine-laser system has been built by K. Hohla et al. employing an oscillator and two amplifier stages. Fig. 23 shows the experimental arrangement. The two amplifiers are operated in a different fashion. The first one may be called an "overshoot" amplifier. This principle can be employed if the time for the build-up of the oscillation is comparable with the time of pumping, that is to say, if very fast pumping is used. In this situation the inversion may transiently exceed the threshold inversion by a considerable amount. For this short time interval the Schaw-

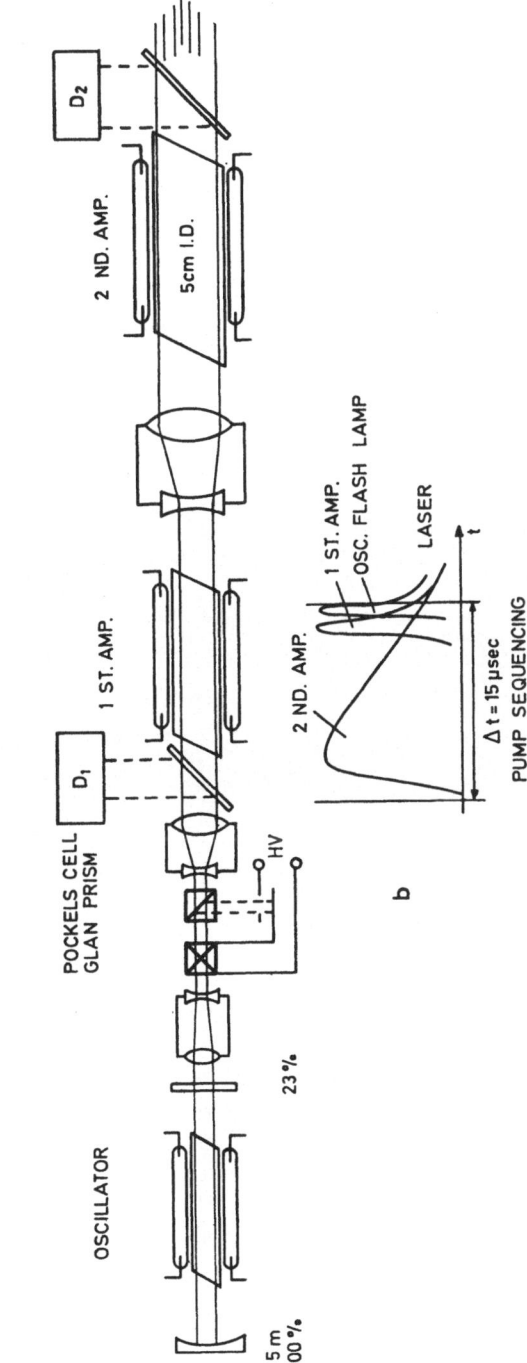

Fig. 23. Experimental set-up of the Gwatt photochemical iodine laser (a). The pulsecutting system after the oscillator consists of a Pockels cell and Glan prism. The Pockels cell is switched by a spark gap which is triggered by the laser light deflected from the prism. Gain can be measured by the diodes D_1, D_2. The duration and sequencing of the flashlamp pulses for pumping the three stages and the switching time of the oscillator are indicated in part (b) of the figure

low-Townes threshold condition which holds for the steady-state oscillator is not applicable. Therefore no optical isolators are needed between the various amplifier stages provided the oscillator pulse is applied before self-oscillation can start in the amplifier chain. This requirement is reflected in the pump sequencing shown in Fig. 23. Fig. 24 shows the considerable pulse sharpening which is found in the first amplifier. The second amplifier is pumped comparatively slowly and the inversion here does not grow above threshold. The pumping of this stage may still be considered fast by comparison with other lasers. This is borne out by the chemical pumping and deactivation requirements discussed in Section 6.1. In addition, shock waves have been found to develop after a while in the photolyzed gas, giving rise to a large disturbance of the beam. This effect also calls for fast operation of the laser.

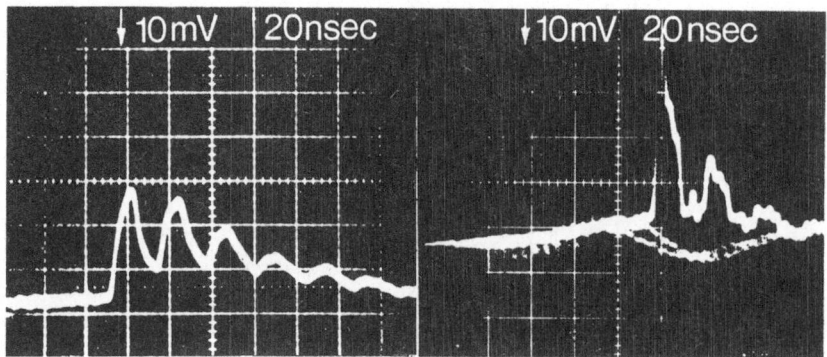

Oscillator signal   Amplified signal

Fig. 24. Pulse-sharpening in the photochemical iodine laser. The left picture shows the oscillator signal, the right one the signal shape after the first amplifier stage of Fig. 23. The series of pulses is due to multiple switching of the Pockels cell. Time scale is 20 nsec/div

It is likely that this type of laser can be sealed up to the 1 $kJ/1$ nsec range. It is to some extent comparable to the neodymium glass laser. Both lasers show similarities with respect to the pumping, the efficiency and the wavelength of emission. Noticeable differences, however, exist for iodine in its comparatively high gain, gain control via linewidth control, and homogeneous broadening of the laser line. Pulses as short as 2 nsec have been measured, although it is not clear whether this measurement was detector-limited. No experimental limitations have appeared so far for the high-power operation of this laser except for the rather low efficiency of 0.5%.

9. Kinetic Information through Chemical Laser Studies

9.1. Characteristics of Oscillator Signals, Computer Modelling

A simple approach is to obtain overall rate constants for the pumping reactions from the pulse shape of the emission signal.

It can be shown [2] that for quasi-steady state conditions of a laser oscillator the intensity of stimulated emission is proportional to the pumping,

$$I(t) \sim P(t) \ . \tag{56}$$

Since this condition is satisfied during the decay of the HF laser emission in certain cases (fast photolysis, low pressure), the time profile of the output intensity reflects the time dependence of the pumping. The pumping term $P(t)$ can be described solely as the formation of the inversion by the reaction as long as losses due to vibrational deactivation are negligible.

With the assumption of negligible vibrational deactivation, the formation of the vibrational inversion is governed by the microscopic rate constants k_v for product formation at the various vibrational levels. The multi-line laser signal, however, results from the total inversion of all lasing P-branch lines which may be written as

$$\sum_{v,J} \Delta N_{v,J}^{v+1,\,J-1} = \sum_{v,J} [N_{v+1,J-1} - (g_{J-1}/g_J)\,N_{v,J}] \tag{57}$$

For exchange reactions of the type $A + BC \rightarrow AB + C$ ($A = F$, $BC = H_2$, HCl, CH_4, C_4H_{10}, D_2) with initial concentrations $A = a$, $BC = b$ and $AB = x = 0$, one may write:

$$P(t) = \frac{d}{dt}\sum_{v,J} N_{v,J}^{v+1,\,J-1} = \sum_{v,J} [k_{v+1,J-1} - (g_{J-1}/g_J)\,k_{v,J}]\,(a-x)\,(b-x)$$

$$= k'(a-x)\,(b-x) \approx k'(a-x)\,b \tag{58}$$

Cascading emission of several vibrational quanta can be accounted for by multiplying k' by a constant factor.

It is indicated in Eq. (58) that the rate equation reduces to first order since, with the density of the F atoms being 10^{-2} Torr and that of the BC reaction partner 0.25 Torr, it follows that $x \ll b$. Thus the rate constants k_{BC} of Table 15 are obtained from the slope of the plots of log I versus time of Fig. 25.

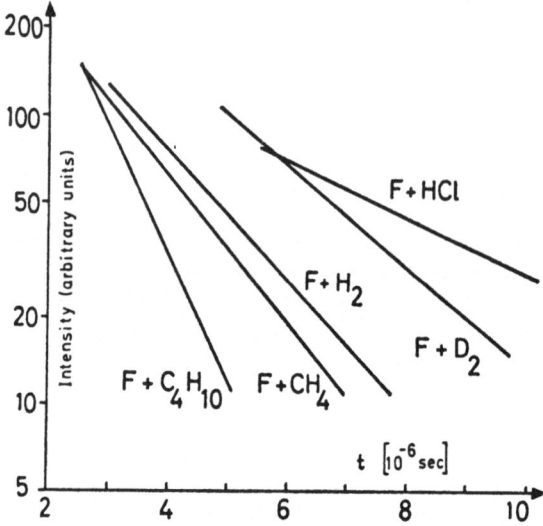

Fig. 25. Decay of HF chemical-laser signals produced in the reactions of F atoms with HCl, H₂, D₂, CH₄ and C₄H₁₀

Table 15. Rate constants of F atom reactions [136)]

| Process | k [cm^{3}mole^{-1} sec^{-1}] | | Relative k | |
	Chemical laser (300 °K $< T$ < 350 °K)	Direct measurement in flow system	Chemical laser	IR luminescence [137)] ($T \sim 350$ °K)
F + H₂	$3.8 \cdot 10^{13}$	$1.6 \cdot 10^{14} \exp(-1600/RT)$ [108)] (300 K $< T <$ 400 °K) $2 \cdot 10^{13}$ ($T = 290$ °K) [107)]	0.88	0.74 ± 0.07
F + D₂	$2.9 \cdot 10^{13}$		0.67	
F + CH₄	$4.3 \cdot 10^{13}$		1	1
F + C₄H₁₀	$7.6 \cdot 10^{13}$		1.77	
F + HCl	$1.5 \cdot 10^{13}$		0.35	0.19 ± 0.02

Other directly measured rate data are available for comparison for the reaction of fluorine atoms with hydrogen only. V.L. Tal'rose *et al.* [107)] find $k_{H_2} = 2 \times 10^{13}$ cm³/mole⁻¹ sec⁻¹ (290 °K) and H. G. Wagner *et al.* [108] give $k_{H_2} = 1.6 \times 10^{14}$ exp-1600/RT. The agreement may be considered satisfactory.

A great deal of more or less detailed computer modelling has been done to predict operational features of chemical lasers since the first studies of this type by Corneil *et al.* [144], Cohen *et al.* [111] and Airey [99]. It is beyond the scope of this review to account for all the computational approaches that have been made. One paper of this kind was reviewed in Section 7 in connection with power predictions for an H_2/F_2 laser oscillator. Here the comprehensive work of Igoshin and Oraevskii [109] on the kinetic processes in an HCl laser may serve as a reference to show the relevant features. The analysis proceeds from the simultaneous solution of chemical kinetics, vibrational relaxation, and radiational processes. The chain reaction model used here is the following

$$Cl_2 + h\nu \longrightarrow 2\ Cl \qquad\qquad Initiation \qquad\qquad (59)$$

$$k_t\,(t) = \gamma_1\,t \cdot 10^{-\gamma_2 t}$$

$\gamma_{1,2}$ are constants which determine duration and intensity of the pumping flash.

$$Cl + H_2 \xrightarrow{\ k_1\ } HCl\,(v = 0) + H \qquad Chain\ stretching \qquad (60)$$

$$k_1 = 8.3 \cdot 10^{13} \exp\,(-5480/RT)\ [cm^3/mole\ sec]$$

$$H + Cl_2 \xrightarrow{\ k_2\ } HCl\,(v = n) + Cl \qquad Laser\ pumping \qquad (61)$$

$$k_2 = 4.1 \cdot 10^{14} \exp(-3000/RT)\ [cm^3/mole\ sec]$$

The probabilities α_n of HCl formation in n-th vibrational state (compare Table 2) were assumed as $\alpha_0 = 0$, $\alpha_1 = 0.134$, $\alpha_2 = 0.482$, $\alpha_3 = 0.362$, $\alpha_4 = 0.022$, $\alpha_5 = 0, (\sum_n \alpha_n = 1)$.

$$H + HCl\,(v = n) \xrightarrow{\ k_3(n)\ } H_2 + Cl \qquad\qquad Inhibition \qquad (62)$$

$$k_3 = 5.9 \cdot 10^{13} \exp(-4500/RT)\ [cm^3/mole\ sec]$$

$$k_3^{(n)}\ for\ n \geqslant 1 = 5.9 \cdot 10^{13}\ [cm^3/mole\ sec]$$

$$Cl + Cl + M \xrightarrow{\ k_t\ } Cl_2 + M \qquad\qquad Chain\ termination \qquad (63)$$

71

The vibrational energy exchange was accounted for by considering reactions of the following type.

$$\text{HCl}(v=n) + \text{HCl}(v=m) \underset{k_{m-1,\,m}^{n+1,\,n}}{\overset{k_{m,\,m-1}^{n,\,n+1}}{\rightleftharpoons}} \text{HCl}(v=n+1) + \text{HCl}(v=m-1) \quad (64)$$

Values for the exchange constants were obtained by calculating the probability $P_{m,\,m-1}^{n,\,n+1}$ of vibrational energy transfer.

$$k_{m,\,m-1}^{n,\,n+1} = z_{mn}\, P_{m,\,m-1}^{n,\,n+1} \quad (65)$$

where z_{mn} is the collision frequency. For calculating the average temperature of the mixture the following thermal equilibrium equation was used:

$$V\varrho\, c_v\,(dT/dt) = W_+ - W_- \quad (66)$$

V is the volume, ϱ is the concentration, c_v is the specific heat at constant volume, W_+ is the rate of heat release by the reaction and W_- is the rate of heat transfer to the walls.

Finally the stimulated processes were included in the following way: the balance equations for the populations N_v, $N_{v'}$ and the photon density $q_{v,\,J}^{v',\,J'}$ in the transition $v', J' - v, J$ having maximum gain were derived as shown in Section 4 of this review.

$$\frac{dN_{v'}}{dt} = -A_{v'}\, _vN_{v'} - \frac{\sigma_v^{v'J'}c}{V}\, q_{v\,J}^{v'J'}\, \Delta_{v\,J}^{v'J'} \quad (67)$$

$$\frac{dN_v}{dt} = A_{v'}\, _vN_{v'} + \frac{\sigma_v^{v'J'}c}{V}\, q_{v\,J}^{v'J'}\, \Delta_{v\,J}^{v'J'} \quad (68)$$

$$\frac{dq_{v\,J}^{v'J'}}{dt} = \frac{\sigma_v^{v'J'}c}{V}\, q_{v\,J}^{v'J'}\, \Delta_{v\,J}^{v'J'} - \frac{q_{v\,J}^{v'J'}}{\tau_p} + A_{v'}\, _vN_{v'J'} \quad (69)$$

V is the resonator volume, c is the speed of light, τ_p is the photon lifetime in the cavity, $\sigma_v^{v'J'}$ is the cross-section for stimulated emission, $A_{v'v}$ is the Einstein coefficient for spontaneous emission, and $\Delta_v^{v'J'}$ is the population inversion. It is assumed that emission occurs in only one vJ-line in a given vibrational band, thus draining the entire population inversion of this band.

The basic results of the calculations are reproduced in Fig. 26. The corresponding experimental parameters are given in the caption to the figure.

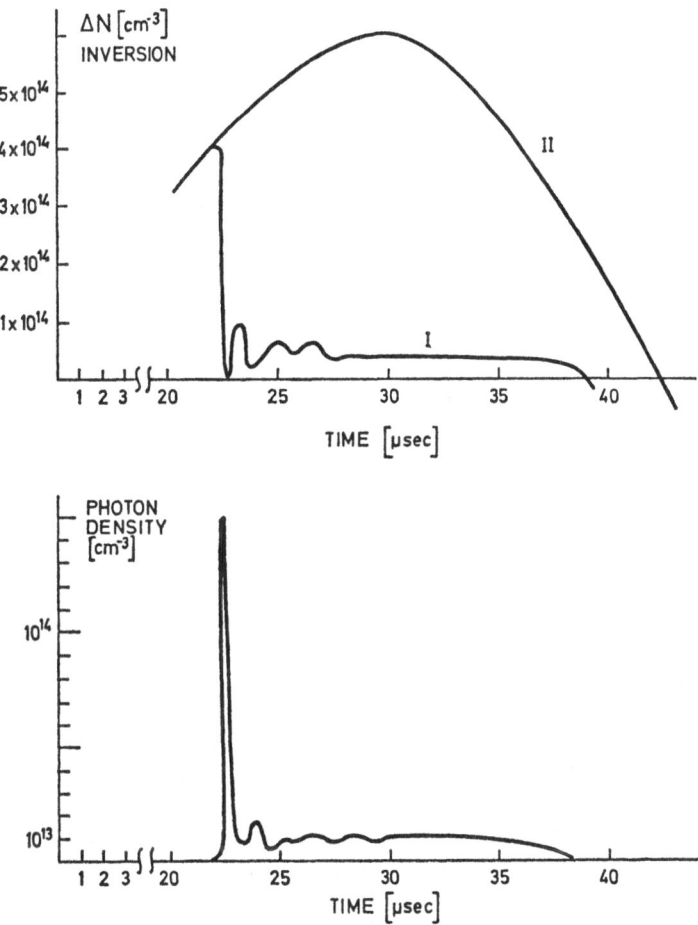

Fig. 26. Calculated time dependence of the inversion in the presence (I) and absence (II) of laser radiation in an HCl laser[109]. The lower picture shows the corresponding emission signal due to $v = 2 \rightarrow 1$ transitions. The parameters were as follows: $\tau p = 1.5 \cdot 10^{-7}$ sec, $\Delta \nu = 0.1$ cm^{-1} atm^{-1}, threshold inversion $\Delta N_0 = 3 \cdot 10^{13}$ cm^{-3}, Cl$_2$: H$_2$ $= 1 : 1$, $2 \cdot 10^{-6}$ mole cm^{-3}, $T_0 = 300$ °K, active volume V and time dependence of the flash, as specified in Ref. 109

Obviously this HCl model is somewhat idealized, and disagreement between calculation and experiment is found in some cases. With even more simplifications and concentration on the essential features, a closed-form solution to such rate equations is possible. This has been attempted, for instance, by Emanuel and Whittier [110]. In their treatment a simplified analysis is presented for intensity, energy, and chemical efficiency of a

73

hydrogen-fluoride chemical laser. Simultaneous effects of laser power extraction, chemical pumping, and chemical deactivation are considered. A multilevel laser is discussed where from lasing threshold to cutoff there is simultaneous lasing between all adjacent vibrational levels. This assumption permits using gain expressions to establish the relative vibrational level populations. The temperature is assumed to be constant and the vibrational energy levels are harmonic. As in other treatments, lasing occurs in only a single J for the entire pulse within one vibrational band. Pumping is provided by the reaction $F + H_2 \rightarrow HF (v = 0,1,2,3) + H$. The rate equations for a chemically reacting system with lasing are then set up in the following form:

$$\varrho \frac{dn(0)}{dt} = \chi_{ch}(0) + \chi_{rad}(0) \tag{70}$$

$$\varrho \frac{dn(1)}{dt} = \chi_{ch}(1) + \chi_{rad}(1) - \chi_{rad}(0) \tag{71}$$

Here $n(v)$ is the number of moles in vibrational level v per unit mass, ϱ is density, $\chi_{ch}(v)$ is net production of $n(v)$ by chemical reactions, $\chi_{rad}(v)$ is production of $n(v)$ by lasing from $v+1 \rightarrow v$. The above equations are inverted for the $\chi_{rad}(v)$.

$$\chi_{rad}(0) = \varrho \frac{d}{dt} n(0) - \chi_{ch}(0) \tag{72}$$

$$\chi_{rad}(1) = \varrho \frac{d}{dt}[n(0) - n(1)] - [\chi_{ch}(0) + \chi_{ch}(1)] \tag{73}$$

If v_f is the highest vibrational level,

$$\chi_{rad}(v_f - 1) = \varrho \frac{d}{dt}[n(0) + \ldots + n(v_f - 1)] \\ - [\chi_{ch}(0) + \ldots + \chi_{ch}(v_f - 1)] \tag{74}$$

A sumation yields

$$\chi_{rad} \equiv \sum_{v=1}^{v_f-1} \chi_{rad}(v) = A_1 - A_2$$

where

$$A_1 \equiv \varrho \sum_{v=0}^{v_f-1}(v_f - v)\frac{dn(v)}{dt} , A_2 \equiv \sum_{v=0}^{v_f-1}(v_f - v)\chi_{ch}(v) \tag{75}$$

The quantity χ_{rad} is the most important figure here since it is used in the determination of energy, intensity and efficiency. The pulse energy $E(t)$ [J/cm^3], for instance, assuming that all transitions have the same wave number ω, per unit volume of gas is given by

$$E(t) = hcN_A \, \omega \int_{t_o}^{t} \chi_{rad} \, dt. \tag{76}$$

χ_{rad} is calculated by assuming gain and threshold conditions and calculating the $n(v) - s$. The chemical kinetics model for the HF laser as usual includes the pumping reaction and deactivation of HF(v) by HF, H$_2$ and F. A comparison with more exact computer solutions shows the validity of this type of analysis. It is considered to be helpful in establishing the relative importance of initial conditions, laser parameters, and rate coefficients for pumping and deactivation reactions.

9.2. Threshold Measurements

If two vibrational rotational lines start to oscillate at the same time, the respective gain products are the same.

$$\sigma \Delta N_{vJ} = \sigma' \Delta N_{v'J'} \tag{77}$$

If the two lines are of the same v and if for a rough approximation the cross-sections σ and σ' are taken to be equal, this means that

$$N_{vJ} - \frac{g_J}{g_{J+1}} N_{v'J+1} = N_{vJ'} - \frac{g_{J'}}{g_{J'+1}} N_{v'\,J'+1} \tag{78}$$

or, more explicitly, if $J' = J + 1$ (neighboring J transitions)

$$(2\,J - 1)\,e^{-(J-1)B_v/T} - (2\,J - 1)\,\frac{N_v}{N_{v'}}\,e^{-J(J+1)\,B_v/T}$$
$$= (2\,J + 1)\,e^{J(J+1)B_v/T} - (2\,J + 1)\,\frac{N_v}{N_{v'}}\,e^{-(J+1)\,(J+2)B_v/T} \tag{79}$$

It follows then that the ratio $N_v/N_{v'}$ is

$$\frac{N_v}{N_{v'}} = \frac{(2\,J - 1)\,e^{-J(J-1)B_v/T} - (2\,J + 1)\,e^{-J(J+1)B_v/T}}{(2\,J - 1)\,e^{-J(J+1)B_v/T} - (2\,J + 1)\,e^{-(J+1)(J+2)B_v/T}} \tag{80}$$

The temperature T in these expressions is the rotational temperature which is equal to the kinetic temperature of the gas. Thus the equal gain

75

condition can be met if the temperature is properly controlled. This so-called equal-gain method has been developed and used extensively by Pimentel and coworkers [112]. A variety of population ratios has been obtained in this way. Examples are quoted in Table 16. Experimental problems

Table 16. Population ratios N_v/N_{v-1} by equal gain (zero gain) measurements

Reaction	N_3/N_2	N_2/N_1	N_1/N_0	Ref.
F + H$_2$ $\longrightarrow$ HF(v) + H		5.5		112)*
F + D$_2$ $\longrightarrow$ DF(v) + D		1.6		112)
1,1$-$C$_2$H$_2$Cl$_2$ $\longrightarrow$ C$_2$HCl + HCl(v)	0.87			138)
	0.85	1.08		113)
cis-1,2$-$C$_2$H$_2$Cl$_2$ $\longrightarrow$ C$_2$HCl + HCl(v)		0.81		138)
	0.71	0.76		113)
trans-1,2$-$C$_2$H$_2$Cl$_2$ $\longrightarrow$ C$_2$HCl + HCl(v)		0.70		138)
{CH$_3$NF$_2$}$^+$ $\longrightarrow$ 2 HF(v) + HCN			0.35	139)
F + CCl$_3$D $\longrightarrow$ DF(v) + CCl$_3$		1.54		154)

* Note added in proof: As a result of more detailed measurements this value has been corrected now and agreement is found with the population ratios of Table 2 of temperature effects are taken into account (G. C. Pimentel, private communication).

arise from the finding that the results appear to be pressure-dependent. For this reason data have to be taken at various pressures and corrections have to be made by extrapolating back to zero pressure. In addition, as Fig. 4 shows, the relative gains become rather insensitive to temperature changes for high total inversion. Consequently the method is most useful for partial inversions. A general problem is apparent in the fact that the dynamics at the start of the oscillation have to be equal and reproducible for the two lines in question. It is most desirable that oscillations should start immediately after threshold is reached and that the laser should attain steady-state behavior. This would require in principle detailed control over cavity losses, gains and cavity modes.

A more sensitive technique developed recently by Pimentel and coworkers [113] is called the "zero-gain technique". Here a tandem laser set-up is used consisting of two laser tubes in a common optical cavity. One contains a known laser system, the "driver". The other mixture, called the "slave", produces the same active reaction product but in an unknown state of vibrational excitation. The onset of oscillations for the driver is then shifted in time, depending on whether the slave produces gain or absorption on certain vibrational rotational lines. Since this effect can also be influenced by varying the rotational temperature, conditions of zero gain — and hence zero population inversion — can be found. This information can be evaluated as explained above for the "equal-gain technique" to yield population ratios N_v/N_{v-1} for the unknown reaction system. It should be mentioned that in laser spectroscopy it is often advantageous to place the probe inside the laser cavity for the measurement of small absorptions [114]. The sensitivity is then increased by a factor which depends on the quality of the Q factor of the cavity and which can be very large (about 10^2 or more), since quite small changes in total absorption may cause large changes in laser intensity, especially if the laser is operated near threshold. Coupled resonators have also been used for such measurements. The first active laser cavity generates the signal whose absorption is to be determined. The probe is placed in a second cavity which is coupled to the first one and which is undamped by an active medium just below the threshold of oscillation. In this way absorption coefficients down to $\alpha \approx 10^{-6}$ have been measured by Boersch and coworkers [115].

All the experimental techniques described here involve the determination of the delay time between initiation of the pumping pulse and laser-pulse onset, or the coincidence of two such delay times belonging to different transitions. An analytical model has been presented by Chester *et al.* [116] to describe the delay τ_c between flashlamp initiation and the start of the laser signal in the flash-photolysis HF chemical laser. The model has been used to predict the functional dependence of τ_c on pressure, flashlamp intensity, optical-cavity losses, and the absolute magnitude of τ_c. However, the possible extension of this work to a detailed vibrational energy-partitioning study has not been demonstrated so far.

9.3. Time-dependent Gain Measurements

The experiments described in the preceding section are based, at least indirectly, on threshold gain determinations. As an extension of this approach, gains have been measured directly.

Laser oscillator-amplifier measurements can be effective in studying the details of the pumping and collisional deactivation processes in chemical

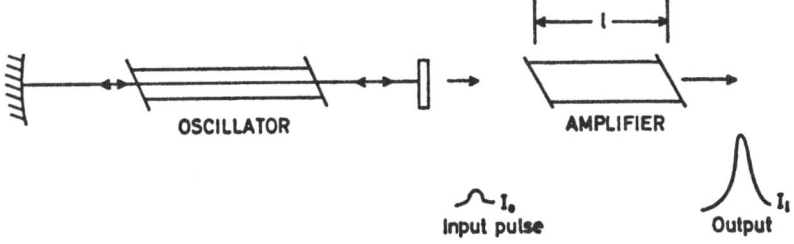

Fig. 27. Schema of an oscillator-amplifier system

laser systems [25]. This technique utilizes the signal from one laser, the oscillator, as a diagnostic aid to study the kinetics of another laser, the amplifier. By operating the oscillator on different molecular transitions and monitoring the change in the signal after it passes through the amplifier, it is possible to study the processes which populate and depopulate the molecular energy levels involved in the laser transition.

An oscillator-amplifier laser system consists of the two principal components shown in Fig. 27. Both operate on the same laser transition. The oscillator provides the input pulse, which is amplified as it passes through the amplifier section. The ratio of the integrated output and input signals is given by [2]

$$\frac{I_l}{I_0} = \frac{h\nu}{\alpha\sigma I_0} \ln \left\{ 1 + (e^{\frac{\sigma\alpha I_0}{h\nu}} - 1)\, e^{\sigma\Delta N} \right\} \tag{81}$$

Here

$$I = \int_0^\tau i_e(t)\, dt$$

is the integrated output intensity (Wattsec/cm^2), I_0 is the integrated input intensity, ΔN is the population inversion per unit area given by

$$\Delta N = (n_2 - \frac{g_2}{g_1} n_1)\, l$$

and $\alpha = 1 + g_2/g_1$. The quantity σ is the cross-section for stimulated emission which is related to the Einstein A coefficient, the wavelength λ, and the linewidth $\Delta\nu$ of the transition by

$$\sigma = \frac{C\lambda^2 A}{8\pi\Delta\nu} \tag{82}$$

Here C is a lineshape factor whose value depends on whether the transition is Doppler- or pressure-broadened.

Fig. 28 shows the variation of I_l with I_0. As can be seen, there are two distinct regions of operation — the small-signal region with

$$\frac{\alpha\sigma I_0}{h\nu} \ll 1$$

and the large-signal region with

$$\frac{\alpha\sigma I_0}{h\nu} \gg 1 \,.$$

In the small-signal region Eq. (81) reduces to

$$I_l/I_0 = e^{\sigma\Delta N} \tag{83}$$

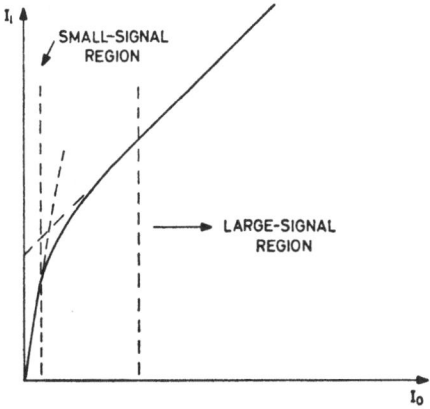

Fig. 28. Variation of amplifier output signal I_l with input signal I_0

and in the large-signal region

$$I_l/I_0 = 1 + \frac{\Delta N h \nu}{\alpha I_0}$$

or

$$I_l = I_0 + \frac{\Delta N h \nu}{\alpha}$$

(84)

Eqs. (83) and (84) provide the basis for obtaining kinetic information from oscillator-amplifier measurements. The amplification is dependent on the population inversion ΔN which depends on the population of the energy levels of the laser transition. Thus, by measuring the changes in the amplification or absorption as the system parameters (temperature of the system, and relative population of the molecular energy levels) are varied, the rates of the interactions affecting the level population can be determined. It should also be mentioned that by combining small- and large-signal gain measurements (Eqs. (83) and (84)), the cross-sections for stimulated emission σ and the linewidths ΔN of the laser transitions can be measured.

For the application of this technique the following requirements have to be met [25]:

(1) The laser transition has to be homogeneously broadened, which is to say that a sufficient number of collisions has to occur during the time of the oscillator pulse. Only then is Δn in (83) the full inversion under the entire line profile.

(2) Linear and non-linear loss processes in the amplifier material must be negligible. This is conditional for the use of Eq. (81). This requirement

79

is normally fulfilled in gas laser materials at low pressures and with sufficiently homogeneous excitation.

(3) In addition, for large-signal gain measurements the energy distribution over the cross-section of the oscillator beam has to be known and must be constant. This condition is much harder to meet than the other requirements.

Combined small- and large-signal gain measurements have been applied to the photochemical iodine laser by Hohla and Kompa [117]. A plot of the gain versus time is shown in Fig. 29 for certain experimental conditions.

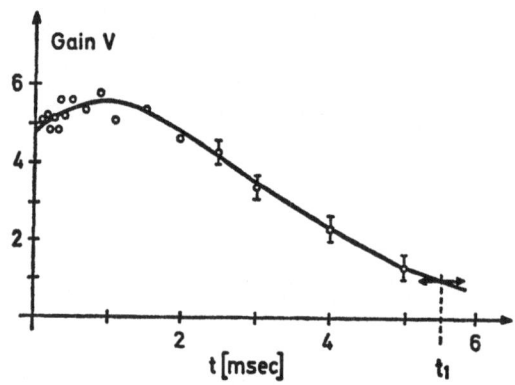

Fig. 29. Plot of the gain $V = \exp. \sigma \Delta N l$ versus time after photolysis at 20 Torr of CF_3I

According to Eq. (83), for small-signal gain the relation $\ln (I_t/I_0) \sim \Delta N$ exists. One may write for the population inversion

$$\Delta n = \frac{\Delta N}{l} = \ln (I_t/I_0) \frac{8\pi\Delta\nu}{\lambda^2 A l C} \tag{85}$$

The lineshape factor $C = .94$ for a Gaussian line. When the measured data of I_t/I_0 are reduced by the use of Eq. (85), the plots of Δn versus t in Fig. 30 are obtained. The plot shows the unexpected result that the population inversion continues to increase after termination of the flash. This behavior is consistent with the observation of several authors that additional chemical pumping processes are operative in the system [44,45], although no definite conclusion on their nature has yet been reached. It can be seen that under the conditions of Fig. 30 the pumping contribution of this reaction is comparable to that of the photodissociation.

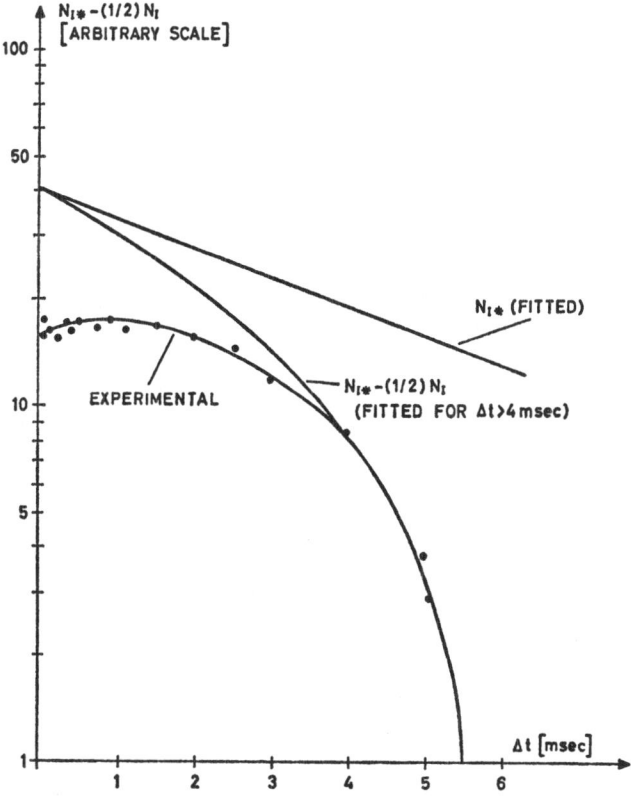

Fig. 30. Decay of the population inversion $\Delta N = N_{\mathrm{I}*} - (1/2) N_{\mathrm{I}}$ and of the concentration of excited iodine atoms $N_{\mathrm{I}*}$ after flash photolysis at 20 Torr of CF_3I

The data of Fig. 30 show that the pumping process has terminated at $t \sim 3$ msec. After that time the decay may be assumed to be controlled by CF_3I deactivation. With this assumption one obtains an expression for the inversion ΔN by introducing $N_{\mathrm{I}*} = x$, $N_{\mathrm{I}} = y$, $N_{CF_3I} = a$, and $x + y = $ const $= x_0$ (no removal of I by I_2 formation).

$$\Delta N = \left(x - \frac{y}{2} \right) = \frac{x_0}{2} \left(3 \exp\left(-\mathrm{kat}\right) - 1 \right) \tag{86}$$

The time t_1 at which the inversion has dropped to $\Delta N = 0$ (or $V = 1$) (Fig. 2) is $t_1 = \ln 3/k\alpha$. Thus one obtains for the rate coefficient k with $a = 20$ Torr CF_3I

$$k = \frac{\ln 3}{t_1 a} = 2.55 \cdot 10^{-16} \left[\mathrm{cm}^3 \ \mathrm{sec}^{-1} \ \mathrm{molecule}^{-1} \right] .$$

81

This is in reasonable agreement with a rate coefficient estimated from flash-photolysis absorption data given by Donovan and Husain [118], $k_{CF_3I} = 4 \cdot 10^{-16}$ [cm^3 sec^{-1} molecule^{-1}]. For the operation of this system as a photochemical laser, it is important to note that under these conditions there are no chemical constraints to the storage of energy for times of several milliseconds.

The measurements can be extended to investigate generally the collisional deactivation of I^* by various added molecules. By comparing small- and large-signal gain measurements, the cross-section for stimulated emission σ was found to be $\sigma = 2 \cdot 10^{-18}$ cm^2 (20 Torr CF$_3$I) and $\sigma = 6 \cdot 10^{-18}$ cm^2 (100 Torr CF$_3$I) [25].

The principle of time-resolved gain spectroscopy was first applied to a molecular chemical laser by L. Henry and coworkers [119]. The HCl laser from the flash photolysis of an H$_2$/Cl$_2$ mixture was chosen for this study. Initial vibrational population figures have been obtained and rate constants derived for the vibrational deactivation, as given in Table 17.

Table 17. Gain spectroscopy, application to H$_2$—Cl$_2$ kinetics

Relative rates of formation k_v of HCl (v)		Decay rates of HCl (v) k_v [sec^{-1} Torr^{-1}]
Polanyi et al [8,9]	Henry et al. [119]	Henry et al. [119]
$v = 0$	0	
1 0.3	0.2—0.28	175
2 0.6	1.1	380
3 1.0	1.0	670
4 0.22	0.29	1100
5 0.03		

Small-signal gain measurements have also been conducted for an HF chemical laser by Gensel et al. [120]. The reaction of fluorine atoms with methane has been used to pump the HF amplifier. Thus the inversion is produced exclusively by

$$F + CH_4 \longrightarrow HF(v) + CH_3 \quad \Delta H = -33.2 \text{ kcal/mole} \qquad (87)$$

This was believed to reduce the chemical complexity of the system. Fluorine atoms to start the reaction are generated in the flash photolysis of tungsten hexafluoride [28]. This provides a relatively clean fluorine source through the photodissociation process (88).

$$WF_6 + h\nu_{<2100A} \longrightarrow WF_n + (6 - n) F \qquad (88)$$

The time behavior of three selected vibrational-rotational inversions is shown in Fig. 31. Some qualitative conclusions may be drawn from this figure. Even at the shortest delay times of $2\,\mu sec$ ($\Delta t_{flash} \sim 2\,\mu sec$) no

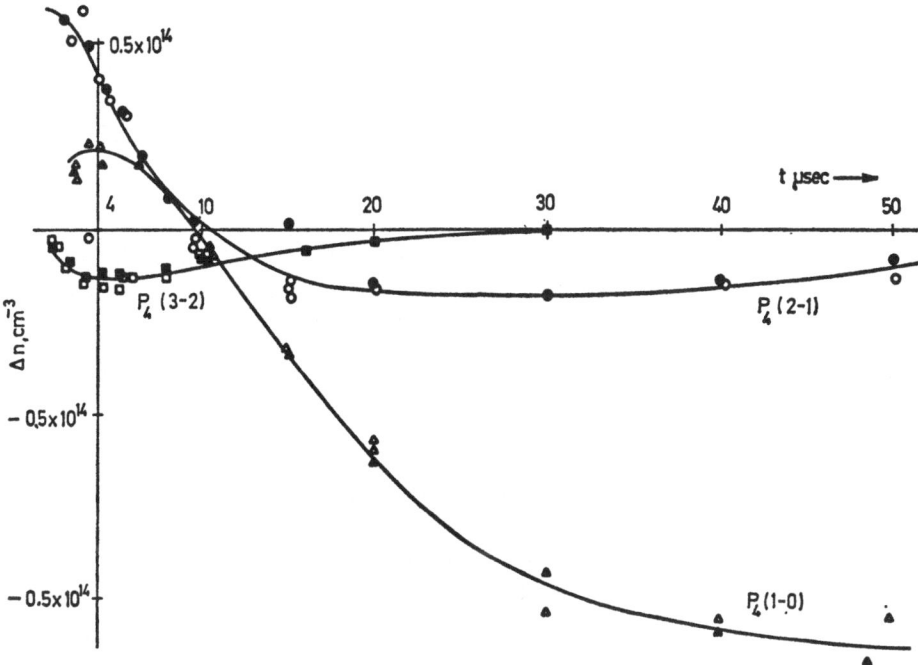

Fig. 31. HF population inversions in WF_6/CH_4 photolysis, $p_{total} = 0.06$ Torr HF. Three vibrational-rotational inversions are shown belonging to transitions in the $v = 3 \rightarrow 2$, $2 \rightarrow 1$ and $1 \rightarrow 0$ bands

positive population inversion is found on $v = 3 \rightarrow 2$ transitions. It is concluded that lasers where such transitions appear (Section 6.6) in the emission are pumped for the most part by the depletion of $n_{v=2}$ through $v = 2 \rightarrow 1$ laser emission. The $v = 2 \rightarrow 1$ lines reach maximum inversion at the end of the flash, while the maximum inversion of the $v = 1 \rightarrow 0$ transitions is seen only at $\Delta t = 4\,\mu sec$. The inversion has completely decayed at $\Delta t = 10\,\mu sec$ under the experimental conditions chosen here. Attempts have been made to interpret the growth and decay of the population inversion on the basis of a computer model using known or in some cases estimated rate constants. The relaxation processes considered to contribute to the deactivation are approximately as listed in Eqs. (49)—(51).

However, agreement between computation and experiment is found only if an additional vibrational-translational deactivation rate is taken into account. It seems uncertain at present whether this excessive deactivation is due to the methyl radicals generated in the system. The role of different chemical compositions should be investigated to identify this deactivation.

A gain-absorption technique has also been used by Smith and coworkers [121] to study the chemical CO laser from the reaction

$$O + CS \longrightarrow CO + S \quad \Delta H = -75 \pm 5 \text{ kcal/mole}$$

Table 18. Relative rates of $O + CS \to CO + S$ into individual vibrational levels of CO [121]

	(a)	(b)	(c)
$v = 4$			
5		0	
6		0.05	
7	≈0.06	0.17	
8	0.27	0.32	
9	0.61	0.41	
10	0.66	0.55	
11	0.80	0.65	≈0.6
12	0.87	0.85	0.87
13	1.0	1.0	1.0
14	0.64	0.90	0.72
15	≈0.2	0.58	0.28
16	0	0.32	0
17	0	0.18	0
18	0	0	0

(a) From infrared emission experiments, $300\,°K < T_{vib}$ (CS) $< 1775\,°K$.
(b) From time-resolved measurements, CS formed by photodissociation of CS_2.
(c) From experiments similar to (b) but with N_2O added, T_{vib} (SC) $\approx 300\,°K$.
All results are quoted relative to $R_{13} = 1.0$.

which occurs as one reaction step in the photolysis of CS_2/O_2 mixtures. A cw CO laser was employed to measure the CO vibrational distribution in this reaction. In the same investigation the more conventional method of spontaneous infrared chemiluminescence measurements was applied to this problem so that a very valuable comparison of the two techniques was obtained which is shown in Table 18. The data confirm the potential of this chemical laser.

Acknowledgements. Where results from our own laboratory have been reported in this paper, I wish to acknowledge with thanks the contributions of P. Gensel (†), K. Hohla, J. R. MacDonald (†), H. Pummer, and J. Wanner.

10. References

[1] Dzhidzhoev, M. S., Platonenko, V. T., Khokhlov, R. V.: Sov. Phys. Uspekhi *13*, 247 (1970). — Basov, N. G., Igoshin, V. I., Markin, J. I., Oraevskii, A. N.: Kvantovaja Elektronika (Russ.) *2*, 3 (1971).

[2] Compare for instance Steele, E. L.: Optical lasers in electronics. New York: J. Wiley and Sons, Inc. 1968. — Röss, D.: Laser, Lichtverstärker und Oszillatoren. Frankfurt: Akademische Verlagsgesellschaft 1966.

[3] Polanyi, J. C.: Appl. Opt. *10*, 1717 (1971).

[4] Herzberg, G.: Molecular spectra and molecular structure I, Spectra of diatomic Molecules. Princeton: D. van Nostrand Co. Inc. 1961.

[5] Polanyi, J. C.: Chemical lasers. Appl. Opt. Supplement *2*, 109 (1965).

[6] Wanner, J.: Dissertation, Universität München 1972.

[7] Vasiliev, G. K., Makarov, E. F., Papin, V. G., Talrose, V. L.: Intern. Symp. Chem. Lasers, Moscow, Sept. 2—4, 1969.

[8] Anlauf, K. G., Kuntz, P. J., Maylotte, D. H., Pacey, P. D., Polanyi, J. C.: Discussions Faraday Soc. *44*, 183 (1967).

[9] Pacey, P. D., Polanyi, J. C.: J. Appl. Opt. *10*, 1725 (1971).

[10] Anlauf, K. G., Maylotte, D. H., Pacey, P. C., Polanyi, J. C.: Phys. Letters *24A*, 208 (1967).

[11] Heydtmann, H., Polanyi, J. C.: J. Appl. Opt. *10*, 1738 (1971).

[12] Johnson, R. L., Perona, M. J., Setser, D. W.: J. Chem. Phys. *52*, 6372 (1970).

[13] Anlauf, K. G., Polanyi, J. C., Wong, W. H., Woodall, K. B.: J. Chem. Phys. *49*, 5189 (1968).

[14] Polanyi, J. C., Tardy, D. C.: J. Chem. Phys. *51*, 5717 (1969). The ratio $E_v/E_{tot} = 0.57$ given in the paper is to be corrected to read $E_v/E_{tot} = 0.67$ (J. C. Polanyi, private communication).

[15] Anlauf, K. G., Charters, P. E., Horne, D. S., MacDonald, R. G., Maylotte, D. H., Polanyi, J. C., Skrlar, W. J., Tardy, D. C., Woodall, K. B.: J. Chem. Phys. *53*, 4091 (1970).

[16] Jonathan, N., Melliar-Smith, C. M., Slater, D. H.: Chem. Phys. Letters *7*, 257 (1970).

[17] Chang, H. W., Setser, D. W., Perona, M. J., Johnson, R. L.: Chem. Phys. Letters *9*, 587 (1971).

[18] Clough, P. N., Polanyi, J. C., Taguchi, R. T.: Can. J. Chem. *48*, 2919 (1969).

[19] Charters, P. E., MacDonald, R. G., Polanyi, J. C.: Appl. Opt. *10*, 1747 (1971).

[20] Jonathan, N., Melliar-Smith, C. M., Slater, D. H.: J. Chem. Phys. *53*, 4396 (1970).

[21] Wagner, H. Gg., Wolfrum, J.: Angew. Chem. *83*, 561 (1971).

[22] Jonathan, N., Melliar-Smith, C. M., Slater, D. H.: Mol. Phys. *20*, 93 (1971).

[23] Flygare, W. H.: Accounts Chem. Res. *1*, 121 (1968).

[24] Airey, J. R., Fried, S. F.: Chem. Phys. Letters *8*, 23 (1971).

[25] Hohla, K.: Dissertation, Universität München 1971.

[26] Furumoto, H. W., Ceccon, H. L.: Appl. Opt. *8*, 1613 (1969).

27) Jonathan, N., Melliar-Smith, C. M., Okuda, S., Slater, D. H., Timlin, D.: Mol. Phys. to be published.

28) Gensel, P., Kompa, K. L., Wanner, J.: Chem. Phys. Letters 7, 583 (1970).

29) Ahlborn, B., Gensel, P., Kompa, K. L.: to be published in J. Appl. Phys.

30) Pummer, H.: Diplomarbeit, Technische Universität München 1972.

31) Compare for instance Merchant, V., Irwin, J. C.: Rev. Sci. Instr. 42, 1437 (1971) and literature quoted there.

32) Chemical lasers. Appl. Opt. Supplement 2 (1965).

33) Kasper, J. V. V., Pimentel, G. C.: Phys. Rev. Letters 14, 352 (1965).

34) Kasper, J. V. V., Pimentel, G. C.: Appl. Phys. Letters 5, 231 (1964). — Kasper, J. V. V., Parker, J. H., Pimentel, G. C.: J. Chem. Phys. 43, 1827 (1965).

35) Schawlow, A. L., Townes, C. H.: Phys. Rev. 112, 1940 (1958).

36) DeMaria, A. J., Ultee, C. J.: Appl. Phys. Letters 9, 67 (1966).

37) Ferrar, C. M.: Appl. Phys. Letters 12, 381 (1968).

38) Gregg, D. W., Kidder, R. E., Dobler, C. V.: Appl. Phys. Letters 13, 297 (1968).

39) Belousova, I. M., Danilov, O. B., Sinitsina, I. A., Spiridonov, V. V.: Soviet Phys. JEPT (English Transl.) 31, 791 (1970).

40) Andreeva, T. L., Dudkin, V. A., Malyshev, V. I., Mikailov, G. V., Sorokin, V. N., Novikova, L. A.: Soviet Phys. JEPT. (English Transl.) 22, 969 (1966).

41) Zalesskii, V. Y., Moskaliv, E. I.: Soviet Phys. JEPT (English Transl.) 30, 1019 (1970).

42) Belousova, I. M., Danilov, O. B., Kladivikova, N. S., Yachnev, I. L.: Soviet Phys.-Tech. Phys. (English Transl.) 15, 1212 (1971).

43) Basov, N. G., Gavrilina, D. K., Leonov, Y. S., Sautkin, V. A.: JEPT Letters (English Transl.) 8, 106 (1968).

44) Andreeva, T. L., Malyshev, V. I., Maslov, A. I., Sobelman, I. I., Sorokin, V. N.: JEPT Letters (English Transl.) 10, 271 (1969).

45) Pollack, M. A.: Appl. Phys. Letters 8, 36 (1966).

46) O'Brien, D. E., Bowen, J.: J. Appl. Phys. 40, 4767 (1969); 42, 1010 (1971).

47) Zalesskii, V. Y., Venediktov, A. A.: Soviet Phys. JETP (English Transl.) 28, 1104 (1969).

48) Deakin, J. J., Husain, D., Wiesenfeld, J. R.: Chem. Phys. Letters 10, 146 (1971). — Husain, D., Donovan, R. J.: Advan. Photochem. 8 (Wiley-Interscience, New York) to be published.

49) Andreeva, I. L., Kuznetsova, S. V., Maslov, A. I., Sobelman, I. I., Sorokin, V. N.: JETP Letters 13, 631 (1971) (Russ.).

50) Velikanov, S. D., Kormer, S. B., Nikolaev, V. D., Sinitsyn, M. V., Solovev, Yu. A., Urlin, V. D.: Soviet Phys. "Doklady" (English Transl.) 15, 478 (1970).

51) Gensel, P., Hohla, K., Kompa, K. L.: Appl. Phys. Letters 18, 48 (1971).

52) De Wolf Lanzerotti, M. Y.: IEEE J. Quant. Electr. QE-7, 207 (1971).

53) Hohla, K., Kompa, K. L.: Chem. Phys. Letters 14, 445 (1972).

54) Hohla, K., Gensel, P., Kompa, K. L.: Proc. 2nd Workshop on Laser Interaction and Related Plasma Phenomena, Rensselaer Polytechnic Institute, Hartford, Connecticut (Plenum Press 1972).

55) Pollack, M. A.: Appl. Phys. Letters 9, 94 (1966).

56) Giuliano, C. R., Hess, L. D.: J. Appl. Phys. 38, 4451 (1967).

57) Pollack, M. A.: Appl. Phys. Letters 9, 230 (1966). — Deutsch, T. F.: Appl. Phys. Letters 9, 295 (1966).

58) Giuliano, C. R., Hess, L. D.: J. Appl. Phys. 40, 2428 (1969).

59) Sorokin, R. P., Lankard, J. R.: J. Chem. Phys. 51, 2929 (1969); 54, 2184 (1971).

60) Campbell, J. D., Kasper, J. V. V.: Chem. Phys. Letters 10, 436 (1971).

References

[61] Shin, H. K.: Chem. Phys. Letters *10*, 81 (1971); J. Phys. Chem. *75*, 1079 (1971).
[62] Ambartzumian, R. V., Apatin, V. M., Letokhov, V. S.: JETP Letters (English Transl.) *15*, 336 (1972). — Ambartzumian, R. V., Letokhov, V. L., Makarov, G. N., Chekalin, N. V.: Chem. Phys. Letters *13*, 49 (1972). — Chen, C. H., Hopkins, B.: 3rd Conf. on Chemical and Molecular Lasers, St. Louis/Mo., May 1972.
[63] Osgood, R. M., Javan, A.: Appl. Phys. Letters *20*, 469 (1972).
[64] Karlov, N. V., Karpov, N. A., Petrov, Yu. N., Prokhorov, A. M., Stelmakh, O. M.: JETP Letters (English Transl.) *14*, 140 (1971). — Basov, N. G., Markin, E. P., Oraevskii, A. N., Pankratov, A. V., Stachkov, A. N.: JETP Letters (English Transl.) *14*, 165 (1971). — Ambartzumian, R. V., Letokhov, V. S.: Appl. Opt. *11*, 354 (1972). — Lyman, J. L., Jensen, R. J.: Chem. Phys. Letters *13*, 421 (1972), for a literature survey see also Kompa, K. L.: Z. Naturforsch. *27b*, 89 (1972).
[65] Patel, C. K. N.: In: Lasers (Ed. Levine, A. K.), Vol. 2, p. 2. New York: Marcel Dekker 1968. — Patel, C. K. N.: Phys. Rev. Letters *12*, 588 (1964); Phys. Rev. *136*, A 1187 (1964). — Patel, C. K. N., Faust ,W. L., MacFarlane, R. A.: Bull. Am. Phys. Soc. *9*, 500 (1964).
[66] For a general reference see for instance IEEE J. Quant. Electr. *QE—8*, No. 2 (1972).
[67] Basov, N. G., Gromov, V. V., Koshelev, E. L., Markin, E. P., Oraevskii, A. N.: JETP Letters (English Transl.) *10*, 2 (1969).
[68] Gross, R. W. F.: J. Chem. Phys. *50*, 1889 (1969).
[69] Pockler, T. O., Shandor, M., Walker, R. E.: Appl. Phys. Letters *20*, 497 (1972). — Basov, N. G., Zavorotnyi, S. I., Markin, E. P., Nikitin, A. I., Oraevskii, A. N.: JETP Letters (English Transl.) *15*, 135 (1972). — Suchard, S. N., Whittier, J. S.: 3rd. Conf. on Chemical and Molecular Lasers, St. Louis/Mo., May 1972.
[70] Dolgov-Savel'ev, G. G., Zharov, V. F., Neganov, Yu. S., Chumak, G. M.: Soviet Phys. JETP (English Transl.) *34*, 34 (1972).
[71] Pummer, H., Kompa, K. L.: Appl. Phys. Letters *20*, 356 (1972).
[72] Polanyi, J. C., Woodall, K. B.: J. Chem. Phys. *56*, 1563 (1972).
[73] Skribanowitz, N., Herman, I. P., Osgood, R. M., Feld, M. S., Javan, A.: Appl. Phys. Letters *20*, 428 (1972).
[74] Suchard, S. N., Pimentel, G. C.: Appl. Phys. Letters *18*, 530 (1971).
[75] Spencer, D. J., Jacobs, T. A., Mirels, H., Gross, R. W. F.: Intern. J. Chem. Kinet. *1*, 493 (1969).
[76] Cool, T. A., Stephens, R. R., Falk, T. J.: Intern. J. Chem. Kinet. *1*, 495 (1969).
[77] Airey, J. R., McKay, S. F.: Appl. Phys. Letters *15*, 401 (1969).
[78] Spencer, D. J., Mirels, H., Jacobs, T. A., Gross, R. W. F.: Appl. Phys. Letters *16*, 235 (1970). — Spencer, D. J., Mirels, H., Jacobs, T. A.: Appl. Phys. Letters *16*, 384 (1970); Opto-Electronics *2*, 155 (1970). — Mirels, H., Spencer, D. J.: IEEE J. Quantum Electron. *QE—7*, 501 (1971). — Spencer, D. J., Mirels, H., Durran, D. A.: J. Appl. Phys. *43*, 1151 (1972).
[79] Meinzer, R. A.: Intern. J. Chem. Kinet. *2*, 335 (1970).
[80] Cool, T. A., Stephens, R. R., Shirley, J. A.: J. Appl. Phys. *41*, 4038 (1970). — Stephens, R. R., Cool, T. A.: Rev. Sci. Instr. *42*, 1489 (1971). — Hinchen, J. J., Banas, C. M.: Appl. Phys. Letters *17*, 386 (1970). — Buczek, C. J., Freiberg, R. J., Hinchen, J. J., Chenansky, P. P., Wayne, R. J.: Appl. Phys. Letters *17*, 514 (1970). — Naegeli, D. W., Ultee, C. J.: Chem. Phys. Letters *6*, 121 (1970). — Glaze, J. A:, Finzi, J., Krupke, W. F.: Appl. Phys. Letters *18*, 173 (1971). — Glaze, J. A.: Appl. Phys. Letters *19*, 135 (1971).
[81] Wittig, C., Hassler, J. C., Coleman, P. D.: Appl. Phys. Letters *16*, 117 (1970). — Suart, R. D., Kimbell, G. H., Arnold, S. J.: Chem. Phys. Letters *5*, 519 (1970). — Jeffers, W. Q., Wiswall, C. E.: Appl. Phys. Letters *17*, 67 (1970). — Suart, R. D., Arnold, S. J., Kimbell, G. H.: Chem. Phys. Letters *7*, 337 (1970).

82) Pilloff, H. S., Searles, S. K., Djeu, N.: Appl. Phys. Letters, in print. — Searles, S. K., Djeu, N.: Chem. Phys. Letters 12, 53 (1971); compare also Foster, K. D., Kimbell, G. H.: J. Chem. Phys. 53, 2539 (1970).

83) Cool, T. A., Falk, T. J., Stephens, R. R.: Appl. Phys. Letters 15, 318 (1969). — Cool, T. A., Stephens, R. R.: J. Chem. Phys. 51, 5175 (1969); Appl. Phys. Letters 16, 55 (1970). — Cool, T. A., Shirley, J. A., Stephens, R. R.: Appl. Phys. Letters 17, 278 (1970). — Brunet, H., Mabru, M.: Compt. Rend. 272B, 232 (1971). — Basov, N. G., Gromov, V. V., Koshelev, E. L., Markin, E. P., Oraevskii, A. N., Shapovalov, D. S., Shcheglov, V. A.: JETP Letters 13, 496 (1971).

84) Cool, T. A.: 3rd Conf. on Chemical and Molecular Lasers, St. Louis/Mo., May 1972.

85) Young, R. A.: J. Chem. Phys. 40, 1848 (1964).

86) Pekar, S. I.: Soviet Phys. "Doklady" (English Transl.) 14, 691 (1970). — Kochelap, V. A., Pekar, S. I.: Soviet Phys. JETP (English Transl.) 31, 459 (1970). — Pekar, S. I., Kochelap, V. A.: Soviet Phys. "Doklady" (English Transl.) 16, 98 (1971).

87) Oraevskii, A. N.: Soviet Phys. JETP (English Transl.) 32, 856 (1971).

88) For instance, Palmer, H. B.: J. Chem. Phys. 26, 648 (1957). — Gorss, R. W. F.: J. Chem. Phys. 48, 1302 (1968). — Gross, R. W. F., Cohen, N.: J. Chem. Phys. 48, 2582 (1968). — Young, R. A., St. John, G. A.: J. Chem. Phys. 48, 895 (1968).

89) For a general reference see Vasil'ev, R. F.: Russ. Chem. Rev. (English Transl.) 39, 529 (1970).

90) For instance Jones, I. T. N., Wayne, R. P.: Proc. Roy. Soc. (London) A 319, 273 (1970). — Evans, W. F. J., Hunten, D. M., Llewellyn, E. J., Vallance Jones, A.: J. Geophys. Res. 73, 2885 (1968).

91) Khan, A. U., Kasha, M.: J. Am. Chem. Soc. 92, 3293 (1970). — Browne, R. J., Ogryzlo, E. A.: Can. J. Chem. 43, 2915 (1965).

92) Dienes, A., Shank, C. V., Trozzolo, A. M.: Appl. Phys. Letters 17, 189 (1970). — Hammond, P. R., Hughes, R. S.: Naval Weapons Ctr. Tech. Publ. 5192 (June 1971).

93) Weller, A.: J. Pure Appl. Chem. 16, 115 (1968). — Rehm. D., Weller, A.: Z. Physik. Chem. NF 69, 183 (1970). — Knibbe, H., Rehm. D., Weller, A.: Ber. Bunsenges. Physik. Chem. 73, 839 (1969). — Weller, A., Zachariasse, K.: Chem. Phys. Letters 10, 646 (1971).

94) Basov, N. G., Danilychev, V. A., Popov, Y. M., Khodkevich, D. D.: JETP Letters (Engl. Transl.) 12, 473 (1970). — Basov, N. G.: Laser Focus 7, 30 (1971).

95) Lorents, D. C.: 3rd Conf. on Chemical and Molecular Lasers, St. Louis/Mo., May 1972.

96) Polanyi, J. C.: Accounts Chem. Res., in print.

97) Hofacker, G. L., Levine, R. D.: Chem. Phys. Letters 9, 617 (1971). — Hofacker, G. L., Michel, K. W.: 3rd Conf. on Chemical and Molecular Lasers, St. Louis/Mo., May 1972.

98) Batalli-Cosmovici, C., Michel, K. W.: Chem. Phys. Letters 11, 245 (1971).

99) Airey, J. R.: J. Chem. Phys. 52, 156 (1970).

100) Kerber, R. L., Emanuel, G., Whittier, J. S.: Appl. Opt. 11, 1112 (1972).

101) Vasil'ev, G. K., Makarov, E. F., Tal'rose, V. L.: JETP Letters (Engl. Transl.) 9, 341 (1969).

102) Pearson, R. K., Cowles, J. O., Hermann, G. L., Pettipiece, K. J., Gregg, D. W.: 3rd Conf. on Chemical and Molecular Lasers, St. Louis/Mo., May 1972.

103) Suchard, S. N., Kerber, R. L., Emanuel, G., Whittier, J. S.: 3rd Conf. on Chem. and Molecular Lasers, St. Louis/Mo., May 1972.

104) Pummer, H., Breitfeld, W., Kompa, K. L., Wedler, G., Klement, R.: Appl. Phys. Letters, in print.

105) Callear, A. B., van den Bergh, H. E.: Chem. Phys. Letters 8, 17 (1971).

106) Pettipiece, K. J.: Chem. Phys. Letters 14, 261 (1972).

References

107) Dodonov, A. F., Lavroskaya, G. K., Morozov, I. I., Tal'rose, V. L.: Dokl. Akad. Nauk SSSR *198*, 622 (1971).
108) Homann, K. H., Solomon, W. C., Warnatz, J., Wagner, H. G., Zetzsch, C.: Ber. Bunsenges. Physik. Chem. *74*, 585 (1970).
109) Igoshin, V. I., Oraevskii, A. N.: Int. Symp. Chem. Lasers, Moscov, September 1969; Khimiya Vysikikh Energii *5*, 397 (1971).
110) Emanuel, G., Whittier, J. S.: Appl. Opt., in print.
111) Cohen, N., Jacobs, T. A., Emanuel, G., Wilkins, R. L.: Inern. J. Chem. Kinet. *1*, 551 (1969).
112) Parker, J. H., Pimentel, G. C.: J. Chem. Phys. *51*, 91 (1969).
113) Molina, M. J., Pimentel, G. C.: J. Chem. Phys. *56*, 3988 (1972).
114) Leite, R. C., Moore, R. S., Whinnery, J. R.: Appl. Phys. Letters *5*, 141 (1964). — Thrash, R. J., v. Weyssenhoff, H., Shirk, J. S.: J. Chem. Phys. *55*, 4659 (1971); compare also Demptröder, W.: Fortschr. chem. Forsch./Topics Current Chem. *17* (1971).
115) Boersch, K., Herziger, G., Weber, H.: Phys. Letters *8*, 109 (1964).
116) Chester, A. N., Hess, L. D.: IEEE J. Quantum Electron. *QE—8*, 1 (1972).
117) Hohla, K., Kompa, K. L.: Z. Naturforsch. *27a*, 938 (1972).
118) Donovan, R. J., Husain, D. J.: Trans. Faraday Soc. *63*, 2023 (1967).
119) Henry, L.: Cambridge Conf. Molecular Energy Transfer, July 1971, Cambridge England. 3rd Conf. Chemical and Molecular Lasers, St. Louis/Mo., May 1972. Menard-Bourcin, F., Menard, J., Henry, L.: To be published.
120) Gensel, P., Kompa, K. L., MacDonald, J. R.: 3rd Conf. on Chemical and Molecular Laser, St. Louis/Mo., May 1972. Max-Planck-Institut f. Plasmaphysik, Lab. Report IV/29.
121) Hancock, G., Morley, C., Smith, I. W. M.: Chem. Phys. Letters *12*, 193 (1971).
122) Kompa, K. L., Pimentel, G. C.: J. Chem. Phys. *47*, 857 (1967). — Kompa, K. L., Parker, J. H., Pimentel, G. C.: J. Chem. Phys. *49*, 4257 (1968).
123) Deutsch, T. F.: Appl. Phys. Letters *10*, 234 (1967).
124) Goldhar, J., Osgood, R. M., Javan, A.: Appl. Phys. Letters *18*, 167 (1971).
125) Pollack, M. A.: Appl. Phys. Letters *8*, 237 (1966).
126) Suart, R. D., Dawson, P. H., Kimbell, G. H.: J. Appl. Phys. *43*, 1022 (1972).
127) Rosenwaks, S., Yatsiv, S.: Chem. Phys. Letters *9*, 266 (1971).
128) Lin, M. C., Brus, L. E.: J. Chem. Phys. *54*, 5423 (1971).
129) Brus, L. E., Lin, M. C.: 3rd Conf. Chemical and Molecular Lasers, St. Louis/Mo., May 1972.
130) Barry, J. D., Boney, W. E., Brandelik, J. E.: Appl. Phys. Letters *.18*, 15 (1971). — Barry, J. D., Boney, W. E., Brandelik, J. E., Mulder, D. M., Woessner, J. K.: Appl. Phys. Letters *20*, 243 (1972); compare also Lin, Y. S., McFarlane, R. A., Wolga, G. J.: Chem. Phys. Letters *14*, 559 (1972).
131) Brandelik, J. E.: Appl. Phys. Letters *19*, 141 (1971).
132) Suchard, S. N., Gross, R. W. F., Whittier, J. S.: Appl. Phys. Letters *19*, 411 (1971).
133) Deutsch, T. F.: Appl. Phys. Letters *11*, 18 (1967).
134) Gregg, D. W., Thomas, S. J.: J. Appl. Phys. *39*, 4399 (1968).
135) Kwok, M. A., Giedt, R. R., Gross, R. W. F.: Appl. Phys. Letters *16*, 386 (1970).
136) Kompa, K. L., Wanner, J.: Chem. Phys. Letters *12*, 560 (1972). — Kompa, K. L., Wanner, J.: To be published.
137) Jonathan, N., Melliar-Smith, C. M., Okuda, S., Slater, D. H., Timlin, D.: J. Mol. Phys., in print.
138) Berry, M. J., Pimentel, G. C.: J. Chem. Phys. *53*, 3453 (1970).
139) Padrick, T. D., Pimentel, G. C.: J. Chem. Phys. *54*, 720 (1971).
140) Stephens, R. R., Cool, T. A.: J. Chem. Phys. *56*, 5863 (1972).

141) Vasil'ev, G. K., Makarov, E. F., Papin, V. G., Tal'rose, V. L.: Soviet Phys. JETP (English Transl.) *34*, 51 (1972).

142) Chang, R. S., MacFarlane, R. A., Wolga, G. J.: J. Chem. Phys. *56*, 667 (1972).

143) Hancock, J. K., Green, W. H.: J. Chem. Phys. *56*, 2474 (1972).

144) Corneil, P. H.: Ph. D. Thesis, Univ. of Calif. Berkeley (1967). — Corneil, P. H.. Kasper, J. V. V.: 2nd Conf. Chem. and Molecular Lasers, St. Louis/Mo., May 1969.

145) Basov, N. G., Kulakov, L. V., Markin, Ye. P., Nikitin, A. I., Oraevskii, A. N.: JETP Letters (English Transl.) *9*, 375 (1969). — Basov, N. G., Galochkin, V. T., Igorkin, V. I., Kulakov, L. V., Markin, Ye. P., Nikitin, A. I., Oraevskii, A. N.: To be published.

146) Hwang, W. C., Kasper, J. V. V.: Chem. Phys. Letters *13*, 511 (1972).

147) Hinchen, J. J.: 3rd Conf. Chemical and Molecular Lasers, St. Louis/Mo., May 1972.

148) Chen, H. L., Moore, C. B.: J. Chem. Phys. *54*, 4072 (1972); see also J. Chem. Phys. *54*, 4080 (1971).

149) Craig, N. C., Moore, C. B.: J. Phys. Chem. *75*, 1622 (1971).

150) Bott, J. F., Cohen, N.: J. Chem. Phys. *55*, 3698 (1971); *55*, 5124 (1971).

151) Solomon, W. C., Blauer, J. A., Jaye, F. C., Hnat, J. C.: Intern. J. Chem. Kinet. *3*, 215 (1971).

152) Bott, J.: 3rd Conf. Chemical and Molecular Lasers, St. Louis/Mo., May 1972.

153) Parker, J. H., Pimentel, G. C.: J. Chem. Phys. *48*, 5273 (1968).

154) Parker, J. H., Pimentel, G. C.: J. Chem. Phys. *55*, 857 (1971).

155) Kompa, K. L., Gensel, P., Wanner, J.: Chem. Phys. Letters *3*, 210 (1969); IEEE J. Quantum Electron. *QE—6*, 185 (1970).

156) Gensel, P., Kompa, K. L., Wanner, J.: Chem. Phys. Letters *5*, 179 (1970).

157) Dolgov-Savel'ev, G. G., Polyakov, V. A., Chumak, G. M.: Soviet Phys. JETP (English Transl.) *31*, 643 (1970).

158) Brus, L. E., Lin, M. C.: J. Phys. Chem. *75*, 2546 (1971).

159) Gross, R. W. F., Cohen, N., Jacobs, T. A.: J. Chem. Phys. *48*, 3821 (1968).

160) Berry, M. J.: 3rd Conf. Chemical and Molecular Lasers, St. Louis/Mo., May 1972.

161) Krogh, O. D., Pimentel, G. C.: J. Chem. Phys. *56*, 969 (1972).

162) Gross, R. W. F., Giedt, R. R., Jacobs, T. A.: J. Chem. Phys. *51*, 1250 (1969).

163) Jensen, R. J., Rice, W. W.: Chem. Phys. Letters *8*, 214 (1971).

164) Gregg, D. W., Krawetz, B., Pearson, R. K., Schleicher, B. R., Thomas, S. J., Huss, E. B., Pettipiece, K. J., Creighton, J. R., Niver, R. E., Pan, J. L.: Chem. Phys. Letters *8*, 609 (1971).

165) Berry, M. J., Pimentel, G. C.: J. Chem. Phys. *49*, 5190 (1968).

166) Berry, M. J., Pimentel, G. C.: J. Chem. Phys. *51*, 2274 (1969).

167) Roebber, J. L., Pimentel, G. C.: 3rd Conf. Chemical and Molecular Lasers, St. Louis/Mo., May 1972.

168) Cuellar-Ferreira, E., Pimentel, G. C.: 3rd Conf. Chemical and Molecular Lasers, St. Louis/Mo., May 1972.

169) Lin, M. C.: J. Phys. Chem. *75*, 3642 (1971); 3rd Conf. Chemical and Molecular Lasers, St. Louis/Mo., May 1972.

170) Spinnler, J. F., Kittle, P. A.: IEEE J. Quantum Electron. *QE—6*, 169 (1970).

171) Burmasov, V. S., Dolgov-Savel'ev, G. G., Polyakov, V. A., Chumak, G. M.: JETP Letters (English Transl.) *10*, 28 (1969).

172) Batovskii, O. M., Vasil'ev, G. K., Makarov, E. F., Tal'rose, V. L.: JETP Letters (English Transl.) *9*, 200 (1969).

173) Hess, L. D.: Appl. Phys. Letters *19*, 1 (1971); J. Chem. Phys. *55*, 2466 (1971).

174) Hess, L. D.: J. Appl. Phys. *43*, 1157 (1972).

175) Florin, A. E., Jensen, R. J.: IEEE J. Quantum Electron. *QE—7*, 472 (1971).

176) Wilson, J., Stephenson, J. S.: Appl. Phys. Letters *20*, 64 (1972).

References

177) Deutsch, T. F.: IEEE J. Quantum Electron. $QE-7$, 174 (1971).
178) Lin, M. C., Green, W. H.: J. Chem. Phys. 53, 3383 (1970).
179) Lin, M. C.: J. Phys. Chem. 75, 284 (1971).
180) Lin, M. C., Green, W. H.: IEEE J. Quantum Electron. $QE-7$, 98 (1971).
181) Lin, M. C., Green, W. H.: J. Chem. Phys. 54, 3222 (1971).
182) Jacobson, T. V., Kimbell, G. H.: Chem. Phys. Letters 8, 309 (1971).
183) Jacobson, T. V., Kimbell, G. H.: J. Appl. Phys. 42, 3402 (1971).
184) Wood, O. R., Burkhard, E. G., Pollack, M. A., Bridges, T. J.: Appl. Phys. Letters 18, 112 (1971).
185) Wood, O. R., Chang, T. Y.: Appl. Phys. Letters 20, 77 (1972).
186) Jensen, R. J., Rice, W. W.: Chem. Phys. Letters 7, 627 (1970).
187) Marcus, S., Carbone, R. J.: IEEE J. Quantum Electron. $QE-7$, 493 (1971).
188) Ultee, C. J.: Rev. Sci. Instr. 42, 1174 (1971); IEEE J. Quantum Electron. $QE-6$, 647 (1970).
189) Wenzel, R. G., Arnold, G. P.: IEEE J. Quantum Electron. $QE-8$, 26 (1972).
190) Pearson, R. K., Cowles, J. O., Herman, G. L., Gregg, D. W.: 3rd Conf. Chemical and Molecular Lasers, St. Louis/Mo., May 1972.
191) Akitt, D. P., Yardley, J. T.: IEEE J. Quantum Electron. $QE-6$, 113 (1970).
192) Coombe, R. D., Pimentel, G. C., Berry, M. J.: 3rd. Conf. Chemical and Molecular Lasers, St. Louis/Mo., May 1972.
193) Kasper, J. V. V., Pimentel, G. C.: Phys. Rev. Letters 14, 352 (1965).
194) Corneil, P. H., Pimentel, G. C.: J. Chem. Phys. 49, 1379 (1968).
195) Deutsch, T. F.: IEEE J. Quantum Electron. $QE-3$, 419 (1967).
196) Airey, J. R.: IEEE J. Quantum Electron. $QE-3$, 208 (1967).
197) Moore, C. B.: IEEE J. Quantum Electron. $QE-4$, 52 (1968); see also Chen, H. L., Stephenson, J. C., Moore, C. B.: Chem. Phys. Letters 2, 593 (1968).
198) Lin, M. C.: Chem. Phys. Letters 7, 209 (1970).
199) Burak, I., Notev, Y., Room, A. M., Szoke, A.: Chem. Phys. Letters 13, 322 (1972).
200) Pollack, M. A.: Appl. Phys. Letters 8, 237 (1966).
201) Arnold, S. J., Kimbell, G. H.: Appl. Phys. Letters 15, 351 (1969).
202) Jacobson, T. V., Kimbell, G. H.: J. Appl. Phys. 41, 5210 (1970).
203) Lin, M. C., Bauer, S. H.: Chem. Phys. Letter 7, 223 (1970).
204) Hancock, G., Smith, I. W. M.: Chem. Phys. Letters 3, 573 (1969).
205) Ahlborn, B., Gensel, P., Kompa, K. L.: J. Appl. Phys., in print.

Received August 11, 1972

W. Demtröder

Laser Spectroscopy

With 16 figures
III, 95 pages. 1971
DM 28,–; US $ 10.40

(Fortschritte der chemischen Forschung Topics in Current Chemistry, Band 17)

With the rapid development of new laser types and the improvement of existing ones, lasers are becoming increasingly important as spectroscopic light sources. This study describes some of the possibilities for applying lasers to modern spectroscopy. As the potential widespread use of the laser becomes more evident, it seems that we are only at the threshold of the "laser revolution". (410 references)

Contents

Spectroscopy with Lasers: Introduction.

Characteristic Features of Lasers as Spectroscopic Light Sources.

Spectroscopic Applications of Lasers.

High-Resolution Spectroscopy based on Saturation Effects.

Spectroscopy of Laser Media.

Conclusions. – Zusammenfassung. –

Prices are subject to change without notice

References.

Springer-Verlag
Berlin Heidelberg New York
München London Paris Sydney Tokyo Wien

In kritischen Übersichten werden in dieser Reihe Stand und Entwicklung aktueller chemischer Forschungsgebiete beschrieben. Sie wendet sich an alle Chemiker in Forschung und Industrie, die am Fortschritt ihrer Wissenschaft teilhaben wollen.

In der Regel werden nur Beiträge veröffentlicht, die ausdrücklich angefordert worden sind. Schriftleitung und Herausgeber sind aber für ergänzende Anregungen und Hinweise jederzeit dankbar. Manuskripte können in den ,,Fortschritten der chemischen Forschung" in Deutsch oder Englisch veröffentlicht werden.

Jeder Band der Reihe ist einzeln käuflich.

This series presents critical reviews of the present position and future trends in modern chemical research. It is addressed to all research and industrial chemists who wish to keep abreast of advances in their subject.

As a rule, contributions are specially commissioned. The editors and publishers will, however, always be pleased to receive suggestions and supplementary information. Papers are accepted for "Topics in Current Chemistry" in either German or English.

Any volume of the series may be purchased separately.